REBEL IDEAS

REBEL IDEAS

THE POWER OF DIVERSE THINKING

MATTHEW SYED

FLATIRON
BOOKS

NEW YORK

www.flatironbooks.com

Designed by Richard Oriolo

Library of Congress Cataloging-in-Publication Data

Names: Syed, Matthew, author.
Title: Rebel ideas : the power of diverse thinking / Matthew Syed.
Description: First U.S. edition. | New York : Flatiron Books, [2021] | Originally published in Great Britain in 2019. | Includes bibliographical references and index.
Identifiers: LCCN 2021002081 | ISBN 9781250769923 (hardcover) | ISBN 9781250769909 (ebook)
Subjects: LCSH: Critical thinking. | Creative thinking. | Divergent thinking. | Thought and thinking. | Problem solving.
Classification: LCC BF441 .S955 2021 | DDC 153.4/2—dc23
LC record available at https://lccn.loc.gov/2021002081

Our books may be purchased in bulk for promotional, educational, or business use. Please contact your local bookseller or the Macmillan Corporate and Premium Sales Department at 1-800-221-7945, extension 5442, or by email at MacmillanSpecialMarkets@macmillan.com.

Originally published in a different form in Great Britain in 2019 by John Murray Press, a Hachette UK company

First U.S. Edition: 2021

10 9 8 7 6 5 4 3 2 1

For Abbas, my inspirational father

CONTENTS

COLLECTIVE BLINDNESS

I

On August 9, 2001, Zacarias Moussaoui, a thirty-three-year-old French Moroccan, enrolled at the Pan Am International Flight Academy in Eagan, Minnesota. This facility, complete with a high-fidelity simulator, provided a comprehensive training program on how to fly commercial airliners. On the surface, Moussaoui seemed like any of the other men who wanted to learn how to fly jumbo jets. He was friendly, inquisitive, and seemingly wealthy. And yet over the course of two days, his instructors became suspicious. He paid for the bulk of the $8,300 course with $100 bills. He seemed unusually interested in the cockpit doors. He kept asking about flight patterns in and around New York.

The staff became so doubtful that two days after Moussaoui enrolled at the school, they reported him to the FBI in Minnesota. He was duly arrested. The FBI questioned him and applied for a warrant to search his apartment, but couldn't show probable cause. Crucially, they failed to connect what they knew about Moussaoui with the broader threat of Islamic extremism. Here was a man with a suspected immigration violation enrolling at a flying school, asking unusual questions, and paying in cash. Weeks later would be the biggest terrorist attack in history.

IN THE MONTHS AFTER 9/11, multiple investigations were launched to work out why such an audacious plot was not foiled by America's intelligence agencies, a group totaling tens of thousands of personnel and in command of a combined budget of tens of billions of dollars. Many of these investigations concluded that the inability to prevent the attack represented a catastrophic failure.

The CIA came in for much of the severest criticism. This is the body, after all, that had been specifically created to coordinate the intelligence community's activities against threats, especially those originating from abroad. From the time the attacks were approved by Osama bin Laden in late 1998 or early 1999, the agencies had twenty-nine months to thwart the plot. They didn't. Richard K. Betts, director of the Saltzman Institute of War and Peace Studies, called it "a second Pearl Harbor for the United States." Milo Jones and Philippe Silberzahn, two leading intelligence experts, described it as "the greatest debacle in the history of the CIA."

One might be tempted to concur, given the clues that had accumulated in the years before 9/11. Al Qaeda had broken its religious taboo on suicide bombing as early as 1993. Bin Laden, a Saudi-born son of a wealthy businessman turned religious zealot, constantly cropped up in raw intelligence reports about Arab terrorist groups. Richard Clarke, a former national coordinator for security under Ronald Reagan, said, "There seemed to be some organizing force and maybe it was he. He was the one thing that we knew the terrorist groups had in common."

Bin Laden publicly declared war on the United States on September 2, 1996, saying in a recorded message that he wanted to destroy the "oppressor of Islam." His strident message was gaining ground among disenfranchised Muslims. Half of terrorist organizations last less than a year, and only 5 percent survive a decade. Al Qaeda had longevity. It was an outlier.

The idea of an airplane being used as a weapon had been circulating for almost a decade. In 1994, an Algerian group hijacked a plane in Algiers and reportedly intended to blow it up over the Eiffel Tower. Later that year, Tom Clancy penned a thriller about a Boeing 747 being flown into the U.S. Capitol Building. It debuted at number one on the *New York Times* bestseller list. In 1995, police in Manila filed a detailed report about a suicide plot to crash a plane into CIA headquarters.

In 1997, Ayman Al Zawahiri—bin Laden's deputy—underscored the intent of Al Qaeda by inciting a massacre of tourists in Egypt, an atrocity that left sixty-two dead, including children. One Swiss woman witnessed her father's head being severed from his body. The Swiss federal police concluded that bin Laden had financed the operation. Unlike previous terrorist groups, Al Qaeda seemed committed to maximizing human suffering, including that of innocents.

In 1998, bin Laden went even further in his thirst for violence against the United States. In a widely published fatwa, he said, "To kill the Americans and their allies—civilians and military—is an individual duty for every Muslim who can do it in any country in which it is possible to do it." On August 7, simultaneous Al Qaeda bombings in Nairobi and Dar es

Salaam killed 224 people and wounded over 4,000. The first was achieved with an explosive device containing more than 2,000 pounds of TNT.

On March 7, 2001, six months before the attack on the World Trade Center, the Russians submitted a report on Al Qaeda providing information on thirty-one senior Pakistani military officers actively supporting bin Laden and describing the location of fifty-five bases in Afghanistan. Soon after, the Egyptian president Hosni Mubarak warned Washington that terrorists were planning to attack President Bush in Rome using an airplane stuffed with explosives. The Taliban foreign minister confided to the American consul general in Peshawar that Al Qaeda was planning a devastating strike on the United States. He feared that American retaliation would destroy his country.

In June 2001, just a few weeks before Moussaoui enrolled at the aviation school in Minnesota, Kenneth Williams, an FBI analyst in Arizona, sent an email to colleagues. It said, "The purpose of this communication is to advise the bureau and NY [New York] of the possibility of a coordinated effort by Osama bin Laden to send students to attend civil aviation universities and colleges." He advised headquarters of the need to make a record of all the flight schools in the country, interview the operators, and compile a list of all Arab students who had sought visas for training. This was to become known as the legendary Phoenix memo. Yet it wasn't acted upon.

With so many pieces of evidence, critics were scathing that the intelligence agencies didn't identify—let alone infiltrate—the plot. The joint senate committee concluded, "The most fundamental problem . . . is our intelligence community's inability to 'connect the dots' available to it before September 11, 2001, about terrorists' interest in attacking symbolic American targets."

It was a damning assessment. Perhaps understandably, the CIA responded defiantly. They defended their record, arguing that it is easy to detect terrorist plots—but only with the benefit of hindsight. They pointed to the research of the psychologists Baruch Fischhoff and Ruth Beyth who, before the historic trip of Richard Nixon to China, asked various people to estimate the probability of different outcomes. Would it lead to permanent diplomatic relations between China and the United States? Would Nixon meet with Mao Zedong at least once? Would Nixon call the trip a success?

The visit was a triumph for Nixon, but what was remarkable was how subjects "remembered" their estimates. Those who thought it would be a disaster recalled being highly optimistic about its success. As Fischhoff put it, "Subjects reconstructed having been less surprised by the events . . . than they really should have been." He called it "creeping determinism."

The 9/11 plot may have seemed glaringly obvious after the event, but was it really so obvious beforehand? Was this not another case of creeping determinism? Was the CIA being condemned for an attack that was, at the time, difficult to detect amid so many other threats?

A nation like the United States is the subject of countless dangers. Terrorist groups stretch around the planet. Surveillance picks up moment-by-moment digital chatter, the vast majority of which amounts to little more than trash talk and idle threats. The agencies could investigate *all* threats, but this would overwhelm their resources. They would be overdiagnosing the problem, hardly an improvement. As one counterterrorism chief put it, the problem was sorting "red flags in a sea of Red flags."

To the CIA and their defenders, 9/11 was not a failure of intelligence but a symptom of complexity. This debate has raged ever since. On one side are those who say that the agency missed obvious warning signs. On the other are those who say that the CIA did everything they reasonably could and that plots are notoriously difficult to detect before the event.

What few people consider is the possibility that both sides were wrong.

II

In the years after it was founded in 1947, the CIA implemented rigorous hiring policies. This was an organization that demanded the best of the best. Potential CIA analysts were put through not only a thorough background investigation, polygraph examination, and financial and credit reviews, but also a battery of psychological and medical exams. And there is no doubt they hired exceptional people.

"The two major exams were an SAT-style test to probe a candidate's intelligence and a psychological profile to examine their mental state," a CIA veteran told me. "The tests filtered out anyone who was not stellar on both tests. In the year I applied, they accepted one candidate for every twenty thousand applicants. When the CIA talked about hiring the best, they were bang on the money."

And yet most of these recruits also happened to look very similar: white, male, Anglo-Saxon, Protestant Americans. This is a common phenomenon in recruiting, sometimes called *homophily*: people tend to hire people who look and think like themselves. It is validating to be surrounded by people who share one's perspectives, assumptions, and beliefs. As the old saying goes, birds of a feather flock together. In their meticulous study of the CIA, Jones and Silberzahn write, "The first consistent attribute of the CIA's identity and culture from 1947 to 2001 is homogeneity of its personnel in terms of race, sex, ethnicity, and class background (relative both to the rest of America and to the world as a whole)." Here is the finding of an inspector general's study on recruitment:

In 1964, the Office of National Estimates [a part of the CIA] had no Black, Jewish, or women professionals, and only a few Catholics. . . . In 1967, it was revealed that there were fewer than twenty African Americans among the approximately twelve thousand nonclerical CIA employees. According to a former CIA case officer and recruiter, the agency was not hiring African Americans, Latinos, or other minorities in the 1960s, a habit that continued through the 1980s. . . . Until 1975, the IC [the U.S. intelligence community] openly barred the employment of homosexuals.[1]

In June 1979, the agency was taken to court for failing to promote female operations officers, settling out of court the following year. A few years later, the agency paid out $410,000 to settle a gender discrimination case brought by an officer with twenty-four years of experience. In 1982, the CIA paid $1 million in a class-action case accusing the agency of the same biases. And yet the CIA didn't significantly alter its personnel policies. "Nothing really changed," one analyst said.

Talking about his experience in the CIA in the 1980s, one insider wrote, "The recruitment process for the clandestine service led to new officers who looked very much like the people who recruited them—white, mostly Anglo-Saxon, middle and upper class, liberal arts college graduates. . . . Few non-Caucasians, few women. Few ethnics, even of recent European background. In other words, not even as much diversity as there was among those who had helped create the CIA."

At a conference in 1999 entitled "U.S. Intelligence and the End of the Cold War," there were thirty-five speakers and presenters, of which thirty-four were white males. "The one exception was a white female who introduced a dinner speaker." Of the three hundred people who attended, fewer than five were not white.

There are no publicly available numbers on the religious orientation of CIA officials responsible for deciding the agency's tasking priorities,

---

[1] This was partly explained by the fear that gay staff, particularly those who had not come out, might be subject to blackmail.

but Jones and Silberzahn state, "We can assume based on what we know of Langley's homogeneity that there were few (if any) Muslims among them." This was corroborated by a former CIA staffer, who said, "Muslims were virtually nonexistent."

Diversity was squeezed further after the end of the Cold War. *Legacy of Ashes*, by the Pulitzer Prize–winning reporter Tim Weiner, quotes Robert Gates, director of the CIA in the early 1990s, as saying that the agency became less willing to employ "people who are a little different, people who are eccentric, people who don't look good in a suit and tie, people who don't play well in the sandbox with others. The kinds of tests that we make people pass, psychological and everything else, make it hard for somebody [with] unique capabilities to get into the agency."

A former operations officer said that through the 1990s, the CIA had a "white-as-rice culture." In the months leading up to 9/11, an essay written for the *International Journal of Intelligence and CounterIntelligence* commented, "From its inception, the intelligence community [has been] staffed by the white male Protestant elite, not just because that was the class in power, but because that elite saw itself as the guarantor and protector of American values and ethics."

The homogeneity at the CIA led to occasional headshaking from politicians who were aware of it. They worried that the CIA was not representative of the society it was created to protect. They believed that if there were more women and ethnic minorities, it would encourage a broader population to feel able to report their concerns, and to imagine working there themselves. They wanted a more inclusive workforce. But CIA insiders always held what seemed like a trump card. Any dilution in their focus on ability, they said, would threaten national security. If you are hiring a sprint relay team, you select the fastest runners. If they are all the same color and gender, so what? To use *any* criteria of recruitment beyond speed is to undermine performance. In the context of national security, they said, putting political correctness above safety is not an acceptable option.

This idea that there is a trade-off between excellence and diversity has a long tradition. In the United States, it formed the basis of a seminal argument by Justice Antonin Scalia for the Supreme Court. Either you can choose diversity, he contended, or you can choose to be "super-duper." If a diverse workforce, student population, or whatever emerges organically through the pursuit of excellence, that is one thing. But to privilege diversity above excellence is different. Diversity above all, the argument goes, is likely to undermine the very objectives that inspired it.

In a relay team, you end up losing the race. If you are a business, it's even worse: you jeopardize your existence. A bankrupt company cannot sustain any workforce, diverse or otherwise. And when it comes to national security, there is a risk that you will imperil the very population you are tasked to protect. And how can that be an ethical course of action? As one former CIA analyst told me, "There was a strong feeling that there should be no compromise. It didn't make sense to 'broaden' the workforce—whatever that means—if it meant that we might lose our cutting edge. It wasn't pigheadedness; it was patriotism."

As late as 2016, security experts were making the same point. In a column for the *National Review*, Fred Fleitz, a former CIA analyst who would become chief of staff for the National Security Council under President Trump, criticized an initiative to increase diversity at the CIA. "Protecting our nation from such threats requires extremely competent and capable individuals to conduct intelligence operations and write analysis in challenging security and legal environments. . . . The CIA's mission is too serious to be distracted by social-engineering efforts."

Part of the reluctance to recruit ethnic minorities was fear of counterespionage, but the skepticism went far deeper. Those who called for a broader intake were shouted down for undermining excellence. The CIA should be about the brightest and the best! Defense is too important to allow diversity to trump ability! As one observer put it, "Political correctness should never be elevated above national security."

What they didn't realize was that this was a false, and perilous, dichotomy.

III

This is a book about diversity. At one level, this might seem like a curious objective. Surely, we should aim to think correctly or accurately, not differently. One should only wish to think differently from other people when they are in the wrong. This seems like common sense.

Another seemingly commonsensical statement, made by Justice Scalia, argued that recruiting people because they are different, in one way or another, is to jeopardize performance. You should hire people because they are smart, or knowledgeable, or fast. Why would you hire people who are less knowledgeable, fast, or talented, just because they are different?

In the coming pages, we will show that both these intuitions are false, at least when it comes to the challenging problems we care most about. If we are intent on tackling our most serious questions, from climate change to poverty, from curing diseases to designing new products, we need to work with people who think differently, not just accurately. And this requires us to take a step back and view performance from a fundamentally different vantage point.

Consider the way we traditionally think about success. If you look at science or, indeed, popular literature, the focus is on individuals. How can we improve the knowledge or perceptiveness of ourselves or our colleagues? Fine books such as *Peak* by Anders Ericsson and Robert Pool, *Sources of Power* by Gary Klein, and *Mindset* by Carol Dweck have become bestsellers. All examine, in their different ways, how we can improve individual abilities through time.

A host of other excellent books follow this approach but in a slightly different way. Even when we have developed expertise, we can still be vulnerable to biases and quirks that undermine our capacity to make wise judgments. *Thinking, Fast and Slow* by Daniel Kahneman, *Predictably Irrational* by Dan Ariely, and *Misbehaving* by Richard Thaler all seek to improve performance by understanding these biases and how to guard against them.

But by focusing on individuals, there has been a tendency to overlook what we might call the *holistic perspective*. Consider a colony of ants. A naive entomologist might seek to understand the colony by examining the ants within the colony. Individual ants, after all, deploy a vast range of behaviors, such as collecting leaves, marching, and so on. They are busy and fascinating creatures. And yet you could spend a year, indeed a lifetime, examining individuals and learn virtually nothing of the colony. Why? Because the interesting thing about ants is not the parts but the whole. Instead of zooming in on individual ants, the only way to understand the colony is to zoom out. One step removed, you can comprehend the colony as a coherent organism, capable of solving complex problems: it builds sophisticated homes, it finds sources of food. An ant colony is an emergent system. The whole is bigger than the sum of its parts.

This book will argue that the same applies to human groups. Pretty much all the most challenging work today is undertaken in groups for a simple reason: problems are too complex for any one person to tackle alone. The number of papers written by individual authors has declined year by year in almost all areas of academia. In science and engineering, 90 percent of papers are written by teams. In medical research, collaborations outnumber individual papers by three to one.

In business, we see the same trend. A team led by Brian Uzzi, a psychologist at Kellogg School of Management, examined more than two million patents issued by the United States since 1975 and found that teams are dominant in every single one of the thirty-five categories. The same trend is seen in the marketplace. Twenty-five years ago, most equity funds were managed by individuals. Now, the vast majority are run by teams. "The most significant trend in human creativity is the shift from individuals to teams, and the gap between teams and individuals is increasing with time," Uzzi writes.

And this is why the holistic perspective is so imperative. We need to think of human performance from the standpoint not of the individual but of the group. From this more rounded perspective, we'll see that diversity is the critical factor driving what we might term *collective* intelligence.

There are, of course, many types of diversity. Differences in gender, race, age, and religion are sometimes classified as *demographic diversity* (or *identity diversity*). We will be focusing instead on *cognitive diversity*. That's to say, variance in perspective, insights, experiences, and thinking styles. There is often (but not always) an overlap between these two concepts. People from different backgrounds, with different experiences, often think about problems differently than one another. We will analyze the precise relationship later in the book.

Cognitive diversity was not so important a few hundred years ago, because the problems we faced tended to be linear, or simple, or separable, or all three. A physicist who can accurately predict the position of the moon doesn't need another opinion to help her do her job. She is already bang on the money. Any other opinion is false. This goes back to our commonsense intuition. Thinking differently is a distraction. With complex problems, however, this logic flips. Groups that contain diverse views have a vast, often decisive, advantage.

Another point worth noting is that these are not speculative claims; rather, they emerge from rigorous, if initially puzzling, axioms. As Scott Page, an expert in complexity science at the University of Michigan, Ann

Arbor, has pointed out, these axioms apply as much to computers as to humans. As we shall see, artificial intelligence today relies on ensembles of algorithms that "think" differently, search differently, and encode problems in diverse ways.

Over the coming pages, the contours of a new science will emerge. Our journey will take us to unusual destinations: the "death zone" at the summit of Mount Everest, the American neo-Nazi movement after the 2008 presidential election, and sub-Saharan Africa at the dawn of our species. We will see why the U.S. Air Force endured so many crashes in the 1940s–50s, how the Dutch reinvented soccer, and why most diets suit almost nobody. We will look at success stories, peeling back the layers of how they happened, and examining their hidden logic. We will look at seminal failures, too. Often, looking at what went wrong provides the most vivid pointers about how to get things right.

By the end of the book, we will be equipped with a fresh perspective on how success happens, one with implications not just for governments and business, but for all of us. To harness the power of cognitive diversity is to gain a key source of competitive advantage, and the surest route to reinvention and growth. We are entering the age of diversity.

But let's start by looking at a selection of puzzles and thought experiments. These will help shed light on what cognitive differences mean and why they matter. We will then return to the buildup to 9/11 and one of the defining intelligence failures of modern times. Often, it is real-world examples that shine the greatest light of all.

IV

In 2001, Richard E. Nisbett and Takahiko Masuda, two social psychologists from the University of Michigan, Ann Arbor, took two groups—one from Japan and the other from the United States—and showed them video clips from underwater scenes. When asked to describe what they had seen, the Americans talked about the fish. They seemed able to recall high levels of detail about the objects. They said things like, "Well, I saw three big fish swimming off to the left; they had white bellies and pink dots." The Japanese, on the other hand, overwhelmingly talked about the context rather than the objects: "I saw what looked like a stream, the water was green, there were rocks and shells and plants on the bottom. . . . Oh, and there were three fish swimming off to the left."

To the experimenters, it was as if the group were seeing different scenes, which they concluded were shaped by differences in culture. America is a more individualistic society; Japanese culture is more interdependent. Americans tend to focus on objects, Japanese on context.

In the next stage of the experiment, the subjects were shown new underwater scenes, with some objects they had seen before and some they had not. When the initial objects were placed in a different context, this threw the Japanese. They struggled to recognize the objects. It was as if the new context diverted their attention. The Americans, on the other hand, had the opposite problem. They were blind to changes in the context.

To the researchers, this was a profoundly surprising result. For decades, a central tenet of psychology was that humans apprehend the world in fundamentally similar ways. This is called *universalism*. As Nisbett put it, "I had been a lifelong universalist concerning the nature of human thought. . . . Everyone has the same basic cognitive processes. Maori herders, Kung hunter-gatherers, and dotcom entrepreneurs all rely on the same tools for perception, memory, causal analysis . . . etc."

The underwater experiment showed that even in our most direct interaction with the world—the act of looking at it—there are systematic differences shaped by culture. Nisbett's paper has now been cited more than a thousand times and inspired a thriving research program. We might say, taking a step back, that Americans and Japanese operate with a different *frame of reference*. The Americans—on average and acknowledging differences within the group—have a more individualistic frame. The Japanese, on the other hand, have a more contextual frame. Each frame attends to useful information. Each frame picks out important features of the underwater scene. Each frame also contains blind spots. The pictures are incomplete.

But now suppose you were to combine a Japanese and an American in a team. Alone, they might perceive only a partial picture. Alone, they would each miss aspects of the scene. Together, however, they would be able to recount both objects and context. By combining two partial frames of reference, the overall picture snaps into focus. You now have a more comprehensive grasp of reality.

THIS EXPERIMENT IS A FIRST, tentative attempt at gently pushing back on one of the intuitions mentioned earlier. You'll remember that Judge Scalia argued that organizations could choose diversity or they could "choose to be super-duper." This implied a trade-off between diversity and excellence. And this is certainly true in a linear task like running (or predicting the orbit of the moon).

And yet the underwater scene experiment hints that, in different contexts, this logic begins to fray. If two people have perspectives that are incomplete, joining them together can yield more insight, not less. They are both wrong, so to speak. They both miss something. But they are wrong in different directions. This means that their shared picture is richer and more accurate. You can glimpse this in a slightly different way by examining a fresh problem, this time something called an *insight puzzle*. Consider the following teaser:

Suppose you are a doctor faced with a patient who has a malignant tumor in his stomach. It is impossible to operate on the patient, but unless the tumor is destroyed, the patient will die. There is a kind of ray that can be used to destroy the tumor. If the rays reach the tumor all at once at a sufficiently high intensity, the tumor will be destroyed. Unfortunately, at this intensity the healthy tissue that the rays pass through on the way to the tumor will also be destroyed. At lower intensities, the rays are harmless to healthy tissue, but they will not affect the tumor, either. What type of procedure might be used to destroy the tumor with the rays and at the same time avoid destroying the healthy tissue?

If you can't solve this puzzle, you are not alone. More than 75 percent of people say that there is no solution and that the patient will die. But now read the following, seemingly unrelated, story:

A fortress was situated in the middle of the country, surrounded by farms and villages. Many roads led to the fortress through the countryside. A rebel general vowed to capture the fortress but learned that mines had been planted on each of the roads. The mines were set so that small bodies of men could pass over them safely, but any large force would detonate them. The general divided his armies into small groups and dispatched each group to the head of a different road. When all was ready, he gave the signal and each group marched down a different road. Each group continued down its road so that the entire army arrived together at the fortress at the same time. In this way, the general captured the fortress.

Now, think back to the medical problem. Can you see the solution now? When tested, more than 70 percent of people found a way to save the patient having read the story about the fortress, triple the initial number. Somehow, by hearing the analogy of the fortress, they were able to glimpse a solution that had previously eluded them. (The solution is to set multiple ray guns around the patient to deliver 10 percent of the radiation with each gun. This destroys the tumor, but without the rays harming healthy tissue.)

This is, of course, an artificial example. But it nevertheless offers a sense of how different perspectives can contribute to solving a challenging problem—in this case, someone with a military background might be of assistance to an oncologist. In such examples, it is not a case of one person being right and another wrong. Rather, it is a case of how looking at a problem through different lenses can jog new insights, new metaphors— and new solutions.

This example challenges intuition in another way, too. When faced with a difficult medical problem, the temptation is to recruit more and more doctors. After all, doctors have the most knowledge when it comes to solving these problems. But if these medical experts bring similar back- grounds and training (and, by implication, similar frames of reference), they are likely to share the same blind spots. Sometimes you need to look at a problem in a new way, perhaps with the eyes of an outsider.

The critical point is that solutions to complex problems typically rely on multiple layers of insight and therefore require multiple points of view. The great American academic Philip Tetlock puts it this way: "The more diverse the perspectives, the wider the range of potentially viable solu- tions a collection of problem-solvers can find." The trick is to find people with perspectives that usefully impinge on the problem at hand.

V

Before resuming our analysis of 9/11, one more key term: *perspective blind- ness*. This term refers to the fact that we are oblivious to our own blind spots. We perceive and interpret the world through frames of reference, but we do not see the frames of reference themselves. This, in turn, means that we tend to underestimate the extent to which we can learn from people with different points of view.

Perspective blindness was the subject of David Foster Wallace's address to Kenyon College in 2005, rated by *Time* magazine as one of the greatest commencement speeches ever given. The speech starts in a fish tank. "There are these two young fish swimming along and they happen to meet an older fish swimming the other way, who nods at them and says, 'Morning, boys. How's the water?' And the two young fish swim on for a bit, and then eventually one of them looks over at the other and goes, 'What the hell is water?'"

Wallace's point is, our modes of thought are so habitual that we scarcely notice how they filter our perception of reality. The danger arises when we overlook the fact that in most areas of life there are other people, with different ways of looking at things, who might deepen our own understanding, just as we might deepen theirs. John Cleese, the British comedian, put it this way: "Everybody has theories. The dangerous people are those who are not aware of their own theories. That is, the theories on which they operate are largely unconscious."

The journalist Reni Eddo-Lodge has offered many examples of perspective blindness. In one, she describes a period when she couldn't afford to take the train all the way to work, so had to cycle part of the way instead. The experience opened a new window on the world:

An uncomfortable truth dawned upon me as I lugged my bike up and down flights of stairs in commuter-town train stations: the majority of public transport I'd been traveling on was not easily accessible. No ramps. No lifts. Nigh-on impossible to access for parents with strollers, or people using wheelchairs, or people with mobility issues, like a walker or a cane. Before I'd had my own wheels to carry, I'd never noticed this problem. I'd been oblivious to the fact that this lack of accessibility was affecting hundreds of people.

This experience provided her with a perspective that she had not merely lacked previously, but didn't know that she lacked. It opened her eyes to a blind spot about her blind spots. This example doesn't imply, of course, that all commuter stations should necessarily be equipped with ramps, stairs, or lifts. But it does show that we can only perform a meaningful cost-benefit analysis if the costs and benefits are perceived. We have to see things before we can make sense of them. This, in turn, hinges on differences in perspective. There are people who can help us see our own blind spots, and whom we can help to see theirs.

Even when we do seek to step beyond our own frames of reference, it turns out to be surprisingly difficult to do so. We can see this in an intuitive way by considering the so-called wedding list paradox. Couples about to get married often register for presents they would love to receive. But what is remarkable is just how often wedding guests depart from the list and purchase a unique gift, a gift they have personally chosen.

Why do guests do this? In 2011, Francesca Gino from Harvard and Frank Flynn from Stanford conducted an experiment to find out. They recruited ninety people and then allocated them to one of two conditions. Half became "senders" while the other half became "receivers." The receivers were then asked to go to Amazon and come up with a wish list of gifts priced between $10 and $30. Meanwhile, the senders were allocated to either choose a gift from the wish list or a unique gift.

The results were emphatic. The senders expected that recipients would prefer unique gifts—ones they had chosen themselves. They supposed that recipients would welcome the personal touch. But they were wrong. Recipients, in fact, much preferred gifts from their own list. The psychologist

Adam Grant reports the same pattern with friends giving and receiving wedding gifts. Senders prefer unique gifts; recipients prefer gifts from their wedding list.

Why? The reason hinges on perspective blindness. Senders find it difficult to step beyond their own frame of reference. They imagine how *they* would feel receiving the gift that they have selected. And, by definition, they would like it a lot, which is why they chose it. Recipients, by contrast, do not experience the anticipated joy, because they have a different set of preferences. Otherwise, they would have put the gift on the list in the first place.

This illustration of how stubbornly static our perspectives often are helps explain why demographic diversity (differences in race, gender, age, class, sexual orientation, religion, and so on) can, in certain circumstances, overlap with cognitive diversity. Teams that are diverse in personal experiences tend to have a richer, more nuanced understanding of their fellow human beings. They have a wider array of perspectives— fewer blind spots. They bridge frames of reference. A study by Chad Sparber, an American economist, found that an increase in racial diversity of one standard deviation increased productivity by more than 25 percent in legal services, health services, and finance. A McKinsey analysis of companies in Germany and the United Kingdom found that return on equity was 66 percent higher for firms with executive teams in the top quartile for gender and ethnic diversity than for those in the bottom quartile. For the United States, the return on equity was 100 percent higher.[2]

Of course, people from the same demographic do not all share the *same* experiences. Black people are not, as a group, homogenous. There is diversity within ethnic groups as well as between them. But this doesn't

---

[2] These studies are suggestive but not yet conclusive. It might not be diversity driving success, but the other way around. The successful firms may be able to afford more diversity. Later, we will bolster the argument that the relationship is causal.

alter the insight that bringing together individuals with contrasting experiences can broaden and deepen the knowledge of the group, particularly when seeking to understand people. This explains another finding, too: homogenous groups don't just underperform; they do so in predictable ways. When you are surrounded by similar people, you are likely not just to share each other's blind spots, but to reinforce them. This is sometimes called *mirroring*. Encircled by people who reflect your picture of reality, and whose picture you reflect back to them, it is easy to become ever more confident of judgments that are incomplete or downright wrong. Certainty becomes inversely correlated with accuracy.

In a study led by Katherine Phillips, professor at Columbia Business School, for example, teams were given the task of solving a murder mystery. They were given plenty of complex material, comprising alibis, witness statements, lists of suspects, and the like. In half the cases, the groups tasked with solving the problem were composed of four friends. The other half were composed of three friends and a stranger—an outsider, someone from beyond their social milieu, with a different perspective. Given what we have learned so far, it should come as no surprise that the teams with an outsider performed better. Much better. They got the right answer 75 percent of the time, compared with 54 percent for a homogenous group and 44 percent for individuals working alone.

But here's the thing. Those in the two groups had very different experiences of the task. Those in diverse teams found the discussion cognitively demanding. There was plenty of debate and disagreement, because different perspectives were aired. The team members typically came to the right decisions, but they were not wholly certain about them. The fact that they had had such a full and frank discussion of the case meant that they were exposed to its inherent complexity.

But what of the homogenous teams? Their experiences were radically different. They found the session more agreeable because they spent most of the time, well, agreeing. They were mirroring each other's perspectives. And although they were more likely to be wrong, they were far more confident about being right. They were not challenged on their blind spots, and so didn't get a chance to see them. They were not exposed to other perspectives, so became more certain of their own. And this hints at the danger with homogenous groups: they are more likely to form judgments that combine excessive confidence with grave error.

VI

Osama bin Laden made his declaration of war on the United States from a cave in Tora Bora in Afghanistan on August 23, 1996. "My Muslim Brothers of the world," he said. "Your brothers in the land of the two holiest sites and Palestine are calling upon you for help and asking you to take part in fighting against the enemy, your enemy: the Israelis and Americans."

Images revealed a man with a beard reaching down to his chest. He was wearing simple cloth beneath combat fatigues. Today, given what we now know about the horror unleashed on the world, the performance of his declaration looks menacing. But here is an insider in the foremost U.S. intelligence agency describing how bin Laden was perceived by the CIA: "They could not believe that this tall Saudi with a beard, squatting around a campfire, could be a threat to the United States of America."

To a critical mass of CIA analysts, then, bin Laden looked primitive and thus of no serious danger to a technological giant like the United States. Richard Holbrooke, one of the most senior officials under President Clinton, put it this way: "How can a man in a cave out-communicate the world's leading communications society?" Another expert close to the CIA said, "They simply couldn't square the idea of putting resources into finding out more about bin Laden and Al Qaeda given that the guy lived in a cave. To them, he was the essence of backwardness."

Now, consider how someone more familiar with Islam would have perceived the very same images. Bin Laden wore simple cloth not because he was primitive in terms of intellect or technology, but because he modeled himself on the Prophet. He fasted on the days the Prophet fasted. His poses and postures, which seemed so backward to a Western audience, were those that Islamic tradition ascribes to the holiest of its prophets. The very images that desensitized the CIA to the dangers of bin Laden were those that magnified his potency in the Arab world.

As Lawrence Wright puts it in *The Looming Tower*, his Pulitzer Prize–winning book about 9/11, bin Laden orchestrated his entire operation by "calling up images that were deeply meaningful to many Muslims but practically invisible to those who were unfamiliar with the faith." This was corroborated by a CIA insider, who said the agency was "misled by the raggedy appearance of bin Laden and his subordinates—squatting in the dirt, clothed in robes and turbans, holding AK-47s, and sporting chest-length beards—and automatically assumed that they are an antimodern, uneducated rabble."

The cave itself had even deeper symbolism. As almost any Muslim knows, Muhammad sought refuge in a cave after escaping his persecutors in Mecca. This was a period known as the *hegira*. The cave was guarded by a series of divine interventions, including an acacia tree that sprouted to conceal the entrance, and a miraculous spider's web and dove's egg that made it seem unoccupied. Muslims know, too, that Muhammad's vision of the Koran occurred in a mountain cave.

To a Muslim, then, a cave is holy. It has deep religious significance. Islamic art overflows with images of stalactites. Bin Laden consciously modeled his exile to Tora Bora as his own personal hegira, and used the cave as a backdrop to his propaganda. As one Muslim scholar and intelligence expert put it, "Bin Laden was not primitive; he was strategic. He knew how to wield the imagery of the Koran to incite those who would later become martyrs in the attacks of 9/11." Wright put it this way: "It was a product of bin Laden's public relations genius that he chose to exploit the presence of the ammunition caves of Tora Bora as a way of identifying himself with the Prophet in the minds of many Muslims who longed to purify Islamic society and restore the dominion it once enjoyed."

The potency of his messages was visible, then, but only to those looking with the right lens. Bin Laden's messages were reaching far and wide, to Saudi Arabia, Egypt, Jordan, Lebanon, Iraq, Oman, Sudan, and even Hamburg, where a group of asylum seekers were radicalized, traveling to Afghanistan in November 1999 at the precise moment the plot to attack Western targets with planes was reaching its culmination in the minds of the Al Qaeda leadership.

The "antimodern, uneducated rabble" had, by now, swelled to an estimated twenty thousand who passed through the training camps between 1996 and 2000, mostly college educated and with a bias toward engineering. Many spoke as many as five or six languages. Yazid Sufaat, who would go on to become one of Al Qaeda's anthrax researchers, had a degree in chemistry and laboratory science from California State University in Sacramento. Many were ready to die for their faith.

Warnings of danger were sprouting from the Muslim world, but the internal deliberations of the CIA discounted them. The CIA were the brightest and the best. They had been hired to analyze threats and to prioritize. Al Qaeda was way down the list, not because the analysts were not studying them intently, but because they couldn't connect what they were seeing.

"The 'beard and campfire' anecdote is evidence of a larger pattern in which non-Muslim Americans—even experienced consumers of intelligence—underestimated Al Qaeda for cultural reasons," Jones and Silberzahn write in *Constructing Cassandra*. A Muslim scholar and expert on U.S. intelligence made the same point: "The CIA couldn't perceive the danger. There was a black hole in their perspective right from the start." Analysts were also misled by the fact that bin Laden had a penchant for issuing his pronouncements in poetry, another point made by Jones and Silberzahn. After the attack on the USS *Cole* in 2000, for example, he hymned lines that included the following:

Sails into the waves flanked by arrogance, haughtiness, and false power,
To her doom she moves slowly. A dinghy awaits her riding the waves,
In Aden, the young men stood up for holy war destroyed,
A destroyer feared by the powerful.

To white, middle-class analysts, this seemed eccentric, almost quaint. Why would you issue orders in verse? It was in keeping with the notion of "a primitive mullah living in a cave." To Muslims, however, poetry has a different meaning. It is holy. The Taliban routinely express themselves in poetry. It is a major aspect of Persian culture. The CIA were studying the pronouncements, but with a skewed frame of reference. As Jones and Silberzahn put it, "The poetry itself was not merely in the foreign language of Arabic; it derived from a conceptual universe light years from Langley."

In the weeks following the USS *Cole* attack, bin Laden's name was scrawled on walls and magazine covers. Tapes of his speeches were sold in bazaars. In Pakistan, T-shirts bearing his photo were sold alongside calendars labeled "Look out America, Osama is coming." Intelligence was picking up broader chatter about a major attack. Words like "spectacular" and "another Hiroshima" were used. The drumbeat leading to 9/11 was now incessant.

Graduates of the camps in Tora Bora had, by now, passed through three stages of military training, with intensive instruction on hijacking, espionage, and assassination. Recruits spent hours studying a 180-page manual entitled *Military Studies in Jihad Against the Tyrants,* which offered state-of-the-art advice on weapons training and infiltration. The pieces were moving ever faster.

The CIA could have allocated more resources to Al Qaeda. They could have attempted infiltration. But they were incapable of grasping the urgency. They did not allocate more resources because they didn't perceive a threat. They didn't seek to penetrate Al Qaeda because they were ignorant of the gaping hole in their analysis. The problem wasn't (just) the inability to connect the dots in the autumn of 2001, but a failure across the entire intelligence cycle. Collaboration should be about broadening and deepening understanding. The homogeneity at the CIA created a vast collective blind spot.

In July 2000, two young men with Arabic names, recently arrived from Europe, enrolled at Huffman Aviation, a flight training school in Florida. Mohamed Atta and Marwan al-Shehhi began their training on a Cessna 152. Ziad Jarrah commenced his training at the Florida Flight Training Center. He was described by his teacher as "the perfect candidate." Hani Hanjour was now engaged in advanced simulator training in Arizona. The endgame was approaching.

Meanwhile, CIA analysts refused to believe that bin Laden was serious about war with the United States. They couldn't recognize the virulence of the germ that had been planted by the leader of Al Qaeda or grasp the significance of the network that he had, by now, erected across the Middle East. Why start a war he couldn't possibly win? That didn't make sense to Western, middle-class analysts. It was another reason why they doubted the prospect of an all-out attack.

They hadn't yet made the conceptual leap—far easier to anyone familiar with an extremist interpretation of the Koran—that victory for the jihadists was to be secured not on earth but in paradise. The code name for the plot among Al Qaeda's inner circle was the Big Wedding. In the ideology of suicide bombers, the day of a martyr's death is also his wedding day, on which he will be greeted by virgins at the gates of heaven. This is their prize, their inspiration.

A daily brief to the president in 1998 mentioned that bin Laden was preparing to hijack planes but didn't discuss the possibility of suicide attacks, instead focusing on the plot to obtain the release of Abdul Basit. The dots depicted a pattern, but a diverse team was required to connect them.

By the summer of 2001, the plot was nearing its culmination. Jordanian intelligence overheard mention of the Big Wedding, and passed on the rumors to Washington, but their significance was not grasped. All of the nineteen hijackers were now within the borders of the United States. In their ears rang the words of bin Laden, later found on the computer of a member of the Hamburg cell: "Wherever you are, death will find you, even in the looming tower." The words were repeated thrice in the speech, an "obvious signal to the hijackers who were on their way."

At almost the same time, senior CIA official Paul Pillar (white, middle-aged, educated at an Ivy League) was discounting the very possibility of a major act of terrorism. "It would be a mistake to redefine counterterrorism as a task of dealing with 'catastrophic,' 'grand,' or 'super' terrorism," he said, "when in fact these labels do not represent most of the terrorism the United States is likely to face or most of the costs that terrorism imposes on U.S. interests."

In their own defense, the CIA pointed to messages and memos that implied an inkling of what was coming, but no reasonable analysis could corroborate this view. The problem at the CIA was not in the details, but the bigger picture. As one intelligence expert put it, albeit in a different context, "It was not so much a matter of particular intelligence reports or even specific policies; instead, it was a deeper intellectual misjudgment of a central historical reality."

On September 10, according to *The Looming Tower*, bin Laden and Al Zawahiri, his deputy, traveled into the mountains above Khost. Their men carried with them a satellite dish and television set so they could watch the atrocities unfold. By this time, the hijackers were in place, prepared and resolute, looking forward to the virgins in paradise.

Jones and Silberzahn speculate that bin Laden "must" have known about the black hole in American intelligence, as on September 9, forty-eight hours before the attacks, he had the "chutzpah to call his mother in Syria and tell her, in effect: 'In two days, you're going to hear big news and you're not going to hear from me for a while.'" The lack of resources allocated to Al Qaeda meant that although the call would be intercepted, the intercept-interpret-analyze cycle for the region was running at a lethargic seventy-two hours. By the time the contents of the call were studied, it would be too late.

At 5 a.m. on the morning of September 11, Atta woke in his room at the Comfort Inn motel at the Portland, Maine, airport. He shaved, got his things together, and then made his way down to reception with Abdulaziz al-Omari, his roommate. At 5:33 a.m. they handed their room key to reception and stepped into a blue Nissan Altima. A few minutes later, they were checking in to U.S. Airways Flight 5930 to Boston, connecting to American Airlines Flight 11 to Los Angeles.

At almost the same time, Waleed and Wail al-Shehri checked out of Room 432 of the Park Inn, in the Newton suburb of Boston, and made their way to Logan International Airport to join Atta. Ahmed and Hamza al-Ghamdi checked out of Days Hotel on Soldiers Field Road, paid for the pornographic film they had purchased, then set off to the airport with their two first-class tickets on United Airlines Flight 175. The other hijackers were also on the move, tickets in their pockets, the Al Qaeda manual on jihad engraved on their minds. *In the plane, as soon as you get on, you should pray to God, because you do this for God, and everyone who prays to God shall prevail.*

None of the hijackers were stopped by security because airport authorities had not been alerted to the threat. The hijackers were permitted to take knives up to four inches in length into the cabin, because intelligence analysts had not grasped the fact that these were the weapons that would enable them to turn jet airliners into deadly missiles.

The first two planes took off just before 8 a.m. At 8:15 a.m., the controller in the Boston Air Traffic Control Center noticed something odd: American Airlines Flight 11 was deviating to the left, over Worcester, Massachusetts, when it should have been turning south. At 8:22, the plane's transponder stopped emitting signals. Six minutes later, the plane banked steeply, as if seeking out the Hudson River Valley. At 8:43, the plane swept over the George Washington Bridge with a deafening roar.

It was now converging with the North Tower like a bullet.

The last thing to do is remember God, and your last words should be that there is no God but Allah and that Muhammad is His prophet. You will notice that the plane will stop and then start flying again. This is the hour in which you will meet God. Angels are calling your name.

VII

The September 11 attacks were a preventable tragedy. Critics of U.S. intelligence were right about that. But the problem was not that the CIA missed obvious warning signs. This is where critics fell prey to creeping determinism, as defenders of the agencies have long claimed. The warning signs were not obvious to the CIA and, ironically, would not have been obvious to many of the groups who sat in judgment over them, many of which themselves lacked diversity. The dearth of Muslims at the CIA is merely one illustration, then, albeit an intuitive one, of how homogeneity undermined the world's foremost intelligence agency. A more diverse group would have created a richer understanding into not just the threat posed by Al Qaeda, but dangers throughout the world, and how different frames of reference, different perspectives, would have created a more comprehensive, nuanced, and powerful synthesis.

A startlingly high proportion of staff at the CIA had grown up in middle-class families, endured little financial hardship, or alienation, or extremism, witnessed few of the signs that act as precursors to radicalization, or had any of a multitude of other experiences that might have added formative insights to the intelligence process. Each would have been assets in a more diverse team. As a group, however, they were flawed. Their frames of reference overlapped. This is not a criticism of white, Protestant, male Americans. It is an argument that white, Protestant, male American analysts—and everyone else—are let down if they are placed in a team lacking diversity.

Consider that the most devastating testimony came from within the ranks of the CIA itself, albeit long after the event. Carmen Medina, a former deputy director who, at the time of her appointment, was one of the few women to make the upper echelons of the agency, fought for diversity during her thirty-two years at Langley, mostly in vain. In a remarkable interview in 2017 for the Cyber Brief, a small digital platform for cyber experts—an interview that barely registered in news circles—she pierced to the heart of one of the greatest intelligence failures in American history. She said:

The CIA has not met its own goals for diversity. If the composition of the U.S. national security community is such that almost everyone has one worldview, we are not in a position to understand our adversaries and anticipate what they are going to do. So, I think it's important that the intelligence community understand and be home to a wide range of views and outlooks about the world.

She went on: "If you really consider differences of opinion and dissenting views and different experiential bases, what you get is a richer and more accurate view of the world."

Perhaps the bitterest irony is that even if the CIA had sensed the warning signs emanating from Afghanistan and beyond, and had decided to infiltrate the Al Qaeda network (the group had operatives in more than twenty-five countries), they would have struggled to do so. Why? Because the lack of diversity among CIA analysts was mirrored by a lack of diversity in the field.

Intelligence expert Milo Jones notes that the CIA had few analysts who could read or speak Chinese, Korean, Hindi, Urdu, Farsi, or Arabic, together making up the languages of more than a third of the world's population. According to the academic Amy Zegart, only 20 percent of the graduating class of clandestine case officers in 2001 were fluent in a non-Romance language. As late as 1998, the CIA didn't employ a single case officer who spoke Pashto, one of the principal languages of Afghanistan. In many ways, this underpinned the mystification of the 9/11 Commission. "The methods for detecting and then warning of surprise attack that the U.S. government had so painstakingly developed in the decades after Pearl Harbor did not fail; instead, they were not really tried." The most expensively assembled intelligence agency in the world never got off the starting blocks.

It is worth noting that many of the TV dramatizations of 9/11 pointed to a different culprit: poor communication between the intelligence services due to interagency rivalry. There were, indeed, many crucial moments, not least a heated meeting between the CIA and FBI in May 2001, when the former refused to disclose information about Khalid al-Mihdhar, who would go on to become one of the five hijackers of American Airlines Flight 77. Some argue that had the CIA shared what they knew, the FBI would have realized that Al Qaeda operatives were already within the borders of the United States.

But while it would be wrong to downplay these and other problems, it would be a mistake to characterize them as the root cause of the failure. The deepest issue was subtler—a problem that existed in plain sight for decades and which Medina put her finger on, albeit too late. Speaking in 2017, she said that the lack of diversity "is such an irony because if any organization needs an effective way of dealing with differences of opinion, it has got to be the intelligence community."

And this is perhaps the greatest tragedy of all. Milo Jones has argued that the failures that characterized the buildup to 9/11 have been repeated throughout the history of the CIA, from the Cuban Missile Crisis and the Iranian Revolution to the failure to anticipate the collapse of the Soviet Union. "Each of these failures can be traced, directly and incontrovertibly, to the same blind spot at the heart of the agency," Jones said when we met in London. And this shows why both sides of this long and sometimes bitter debate—both those who defended the intelligence services and those who attacked them—have overlooked the key issue. The detractors were right to say that the threat was obvious in hindsight. The defenders were right to respond that the CIA hired highly talented people, and the threat was not obvious to them.

What is certain is that no blame should be attached to any individual analyst. They were not lazy, or asleep on the job, or negligent, or any of the accusations typically used to explain underperformance. They didn't lack insight or patriotism or work ethic. Indeed, it could be argued that no single intelligence analyst lacked anything at all. What they lacked emerged only at the level of the group.

The CIA were individually perceptive but collectively blind. And it is in the crosshairs of this paradox that we glimpse the imperative of diversity.

REBELS
VERSUS
CLONES

I

Take a look at the following crossword. There are thirty-five clues in total, eighteen across and eighteen down. Some of the clues are general knowledge, some are riddles, and others are anagrams. If you want to have a go at it, the answers are at the back of the book. This particular crossword was published in the *Daily Telegraph* on January 13, 1942. At the time, readers of the newspaper had been complaining that the daily crossword puzzle was becoming too easy. Indeed, some claimed they could complete it in a matter of minutes. This was met with disbelief in some quarters, prompting a man called W. A. J. Garvin, chairman of the Eccentric Club,

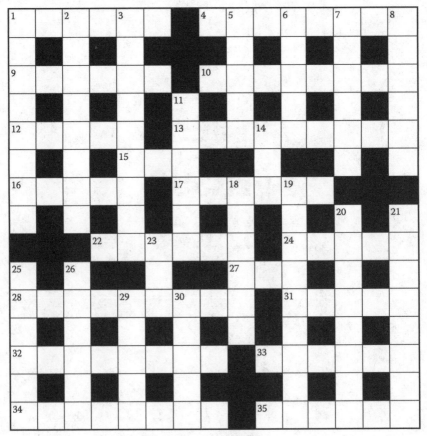

Crossword 5,062
Daily Telegraph, January 13, 1942

to offer a £100 prize to be paid to charity if anyone could complete the puzzle in less than twelve minutes.

Arthur Watson, then editor of the *Daily Telegraph*, responded by hosting a competition for anyone who thought they could meet Garvin's challenge. On January 12, more than thirty people traveled to the newsroom on Fleet Street to tackle the crossword under controlled conditions. The crossword then appeared in the next day's edition.

REBELS VERSUS CLONES

ACROSS

1. A stage company (6)
4. The direct route preferred by the Roundheads (5,3)
9. One of the evergreens (6)
10. Scented (8)
12. Course with an apt finish (5)
13. Much that could be got from a timber merchant (5,4)
15. We have nothing and are in debt (3)
16. Pretend (5)
17. Is this town ready for a flood? (6)
22. The little fellow has some beer; it makes me lose color, I say (6)
24. Fashion of a famous French family (5)
27. Tree (3)
28. One might of course use this tool to core an apple (6,3)
31. Once used for unofficial currency (5)
32. Those well brought up help these over stiles (4,4)
33. A sport in a hurry (6)
34. Is the workshop that turns out this part of a motor a hush-hush affair? (8)
35. An illumination functioning (6)

DOWN

1. Official instruction not to forget the servants (8)
2. Said to be a remedy for a burn (5,3)
3. Kind of alias (9)
5. A disagreeable company (5)
6. Debtors may have to do this money for their debts unless of course their creditors do it to the debts (5)
7. Boat that should be able to suit anyone (6)
8. Gear (6)
11. Business with the end in sight (6)
14. The right sort of woman to start a dame school (3)
18. "The war" (anag.) (6)
19. When hammering take care not to hit this (5,4)
20. Making sound as a bell (8)
21. Half a fortnight of old (8)
23. Bird, dish, or coin (3)
25. This sign of the zodiac has no connection with the fishes (6)
26. A preservative of teeth (6)
29. Famous sculptor (5)
30. This part of the locomotive engine would sound familiar to the golfer (5)

One of the contestants that afternoon was Stanley Sedgewick, a clerk with a firm of city accountants. He had become something of a crossword whiz during his daily train journey to work. "I became quite good at solving the crossword puzzles appearing in the *Daily Telegraph*," he later said. "I went along to find about thirty other would-be fast solvers. We sat at individual tables in front of a platform of invigilators including the editor, Mr. Garvin, and the timekeeper."

In the event, four of the contestants solved the puzzle in time and although Sedgewick was a word short when the bell rang, he impressed observers with his ingenuity and lateral thinking. The contestants were then treated courtesy of the *Daily Telegraph*. "We were given tea in the chairman's dining room, and dispersed with the memory of a pleasant way of spending a Saturday afternoon," Sedgewick said.

Several weeks later he received a letter, in an envelope marked "Confidential." This was not the kind of thing accounting clerk Stanley Sedgewick got in the mail. Intrigued, he picked it up.

"Imagine my surprise," he said, "when I received a letter . . . inviting me, as a consequence of taking part in the *Daily Telegraph* Crossword Time Test, to make an appointment to see Colonel Nicholls of the General Staff who would 'very much like to see you on a matter of national importance.'"

The world was at war. The previous year, Hitler had commenced Operation Barbarossa, the invasion of Russia, and the United Kingdom was in a state of acute vulnerability. And at Bletchley Park, an estate in rural Buckinghamshire, fifty miles northwest of London, a team of men and women were assembled to work on the most secret of missions.

The Enigma machine was an encryption device used by Nazi Germany across all branches of its armed forces. The device was small, not dissimilar to a typewriter in a wooden box, with an encryption technology that consisted of an electromechanical rotor mechanism that scrambled the twenty-six letters of the alphabet. An operator would enter text on the keyboard and another would write down which of twenty-six lights above the keyboard illuminated with each key press. Many in German high command thought that this encryption method was unbreakable.

The group at Bletchley Park was recruited by Britain's secret intelligence services to try, among other things, to crack the Enigma. The site was a mansion in "an ugly mix of mock-Tudor and Gothic styles, built in red brick and dominated on one side by a large copper dome turned green by exposure to the elements," according to Michael Smith in his superb book *The Secrets of Station X*. Much of the work took place in temporary wooden huts constructed on the grounds.

Although these huts were rudimentary, they would play host to some of the most important (and fascinating) activity of the Second World War. The Bletchley Park team cracked the Enigma, providing a treasure trove of information that would prove vital to the overall war effort. Some argue that the intelligence shortened the war by up to three years. Others claim that it altered the outcome itself. Winston Churchill would describe Bletchley Park as "the goose that laid the golden egg."

Now, if you were recruiting a crack team of code breakers, I am guessing you would want to hire world-class mathematicians. This was precisely the approach of Alistair Denniston, a diminutive Scot, when he was asked to head up the Bletchley Park operation. In 1939, he hired Alan Turing, then a twenty-seven-year-old fellow at King's College, Cambridge, and Peter Twinn, a twenty-three-year-old from Brasenose College, Oxford. Over time, more mathematicians and logicians would be added to the team.

These recruits would prove crucial, and the vast majority of historical analysis of Bletchley Park, and its outsize contribution to winning the war, has focused on their remarkable minds. The narrative told in articles, books, and films holds that they simply outthought their opponents, developing the breakthroughs that changed everything.

These individuals were, indeed, brilliant. Turing was one of the great mathematical geniuses of the twentieth century. Yet they would have been incapable of solving this complex, historically significant set of problems on their own—no more capable than the homogenous analysts at the CIA who failed to spot the 9/11 plot. No, the Bletchley Park miracle hinged on something extra: on people we rarely read about and insights that we rarely celebrate. This extraordinary team succeeded because it was optimized for cognitive diversity.

In this chapter, we are going to dig deeper into this critical concept. Why do homogenous institutions tend to fail, often without realizing why? How and why do diverse teams become more than the sum of their parts? We will see why diversity is coming to the fore across all branches of science, not to mention the strategies of cutting-edge institutions.

And we will return to Stanley Sedgewick and see how a lowly bank clerk made a critical contribution to the tide of modern history.

II

We can express the basic idea of diversity science in visual form. Suppose the rectangle in figure 1 represents the universe of useful ideas, which is to say, the insights, perspectives, experiences, and thinking styles relevant to a particular problem or objective. We might call this the *problem space*.

With simple problems, one person might possess all this information. Diversity is unnecessary. But with complex problems, no one person will have all the relevant insights. Even the smartest individual will have only a subset of knowledge. We can represent a smart person, David, with the circle. He knows a lot, but he doesn't know everything.

We can now see the dangers of homophily. In figure 2, we can see

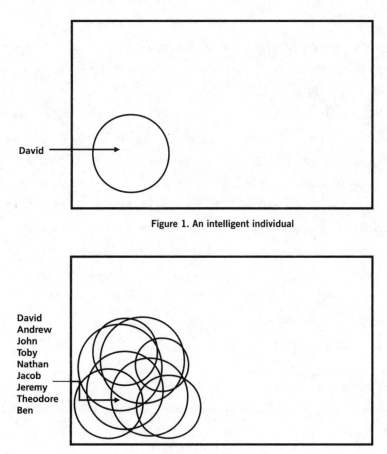

Figure 1. An intelligent individual

Figure 2. An unintelligent team

what happens when a group of people come together who think in the same way. Every individual is smart. They each have impressive knowledge. But they are also homogenous. They know similar things and share the same perspectives. They are, so to speak, "clone-like." This, of course, was the basic problem at the CIA.

Think how comforting it is to be surrounded by people who think in the same way, who mirror our perspectives, who confirm our prejudices. It makes us feel smarter. It validates our worldview. Indeed, evidence from brain scanners indicates that when others reflect our own thoughts back to us, the pleasure centers of our brains are stimulated. Homophily is somewhat like a hidden gravitational force, dragging human groups toward one corner of the problem space.

These dangers are as ancient as mankind itself. They were certainly well understood by the Ancient Greeks. In *Nicomachean Ethics* Aristotle writes that people "love those who are like themselves." Plato notes in *Phaedrus* that "similarity begets friendship." The phrase "birds of a feather flock together" is derived from the early pages of Book I of Plato's masterpiece *The Republic*. Indeed, if you look closely enough, the danger of intellectual conformity is an abiding preoccupation of Ancient Greek culture. And this is why figure 2 is worth keeping in mind, for it represents a pervasive problem in the world today: groups of people who are individually intelligent but collectively, well, stupid.

THE POLL TAX, or community charge, has been surrounded by infamy since it was first introduced by the British government in the late 1980s. The centerpiece of the policy was a change in local taxation from a levy charged on property to one paid by individuals (specifically, a flat per-capita tax for every adult, the amount set by local authorities). It had numerous defects that should have killed it off at birth. The tax was almost impossible to collect. It was impractical to deliver. It was also regressive, falling disproportionally on households in modest, low-valued dwellings.

Some households would prove to be more than £1,500 worse off when the policy was enacted. Many more were at least £500 worse off. In 1989, this represented a substantial proportion of combined household income. Meanwhile others, a small minority, were no less than £10,000 a year better off. This inequity would have consequent effects. Protests were inevitable, exacerbating the inherent difficulties of collecting the tax. Nonpayment was pretty much baked into its conception. As the tax was implemented, the results were as predictable as they were disastrous. As one source put it, "The burden of collecting the tax precipitated a virtual collapse in the finances of some city authorities."

Peaceful protesters took to the streets shouting, "Can't pay, won't pay!" Some marches were infiltrated by militants, and rioting ensued. A march through London with up to a quarter of a million people led to the smashing of shop windows, cars set ablaze, and stores looted. In total, there were 339 arrests and more than 100 injuries. For a few heady days, people feared that the violence might prove contagious.

Anthony King and Ivor Crewe, two experts on British politics, write:

> More than two decades later the whole episode still evokes wonder and astonishment. . . . Every dire prediction made about the poll tax was sooner or later fulfilled. Its perpetrators walked into clearly visible traps with their eyes open, but they evidently saw nothing. They blundered on, impervious to warnings. In the end, their failure was abject and total.

How did this debacle happen? According to King and Crewe, the poll tax is part of a deeper pattern that extends across postwar British political history. They argue that for all their superficial differences, "a substantial proportion" of all the biggest blunders, by governments of all political complexions, share the same root cause: a lack of diversity. In particular, they focus on the lack of social diversity in political elites. In the case of the poll tax, they note that Nicholas Ridley, the secretary of state for the environment responsible for its implementation, was the son of literal aristocrat Viscount Ridley and grew up in a magnificent estate known as Blagdon Hall. His mother was a daughter of the architect Sir Edwin Lutyens and a niece of the painter Neville Lytton. The other environment secretaries during the life of the poll tax were Patrick Jenkin, Kenneth Baker, and Chris Patten, all of whom went to elite private schools and then either Oxford or Cambridge. (Ridley, for his part, went to Eton and Oxford.)

As for the review group, this was led by William Waldegrave, whose father was—deep breath—Geoffrey Noel Waldegrave, Twelfth Earl Waldegrave, KG, GCVO, TD, known as Viscount Chewton, Lord Warden of the Stannaries. His mother was Mary Hermione Grenfell, herself descended from a line of imperial businessmen. He grew up in Chewton House, one of the largest mansions in the county of Somerset.

In his memoir, *A Different Kind of Weather,* Waldegrave is admirably honest about just how detached he was from the lives of most people. "I never played with a local child," he writes. "When we mentioned our neighbors, we meant the Jolliffes at Ammerdown, eight miles away, or the Asquiths at Mells, the Duckworths at Orchard Leigh, the Hobhouses at Hadspen, or the Bishop at his palace at Wells."

The Waldegraves holidayed at Loch Moidart with others of their class, where a famous concert pianist would play the grand piano in the drawing room. Their other holiday destination was Champéry, where they took a horse-drawn sleigh to their chalet, the Chalet des Frênes. The young Waldegrave had a cook and a governess. He and his brother shot pheasants on their estate. When a Black man was seen near the house, his mother's first thought was that he was a terrorist, and she grabbed an edge trimmer to confront him. She then realized that the "terrorist" had gathered up young William, who had crashed his bicycle out of sight, and was trying to help him.

According to King and Crewe, Waldegrave may have had a background very different from the general population, but it was similar to other members of the review team. Not all were as privileged as Waldegrave himself, but they each came from unusually wealthy backgrounds.

"No member represented any other section of British society," they note. This was a group squeezed not just into a single quadrant of the problem space, then, but into the tiniest of pixels. They were smart, but they were also homogenous. They were not clones in a genetic sense, but in a demographic sense. And when it comes to politics, where diversity of experience is so critical to informing policy choices, this would prove catastrophic.

And yet, the review group *loved* working together. King and Crewe quote insiders talking about a "remarkable esprit de corps." They were agreeing, mirroring, parroting, corroborating, confirming, reflecting. They were basking in the warm glow of homophily. This social harmony deluded them into thinking they were homing in on a wise policy. In fact, it showed the opposite. They were entrenching each other's blind spots.

As they played bridge together, and even went to parties together, they couldn't hear the alarm bells that would have been deafening to anyone familiar with diversity science. And is it any wonder that such a clone-like group struggled to foresee the practical problems of collection, the difficulties of implementation, and the fact that families would struggle to pay? Is it any wonder that they failed to recognize the pressure this would put on local government and, ultimately, the social fabric itself?

For elderly people, the cost could be particularly devastating: "A pensioner couple in inner London could find themselves paying 22 percent of their net income in poll tax, whereas a better-off couple in the suburbs pay only 1 percent." Yet when confronted with the tragic situation facing elderly couples who didn't have the cash to pay, Nicholas Ridley struggled to grasp the problem. He replied (apparently seriously), "Well, they could always sell a picture."

A few years earlier, Patrick Jenkin, Ridley's predecessor, made a similarly revealing comment during the energy crisis of the 1970s. In a TV interview, he encouraged the public to save electricity by "brushing their teeth in the dark." It later emerged that Jenkin himself used an electric toothbrush, and his north London home was photographed with lights on in every room.

The problem with the poll tax wasn't, according to King and Crewe, with any individual politician or official. Many were devoted public servants and would go on to have distinguished careers. They were also impressive thinkers. The psephologist Sir David Butler quoted an insider as saying that they were "the brightest selection of people ever gathered" to consider local government reforms. King and Crewe also point out that privilege should never be a bar to high office, and that many from wealthy backgrounds, inherited or otherwise, have contributed much to the common good.

But this cuts to the essence of the problem: when smart people from a singular background are placed into a decision-making group, they become collectively blind. As King and Crewe put it, "Everyone projects onto others his or her lifestyles, preferences, and attitudes. Some do it all the time; most of us do it some of the time. [Officials in] Whitehall and Westminster unthinkingly project onto others values, attitudes, and whole ways of life that are not remotely like their own."[3]

This isn't just about Conservative politicians, of course. King and Crewe cite many examples involving Britain's more left-wing Labour Party. One instance was in a speech in July 2000 by Tony Blair, who called for new powers for the police to deal with antisocial behavior. "A thug might think twice about kicking in your gate, throwing traffic cones around your street, or hurling abuse into the night sky if he thought he might be picked up by the police, taken to a cashpoint, and asked to pay an on-the-spot fine of, for example, £100," he said. The response to this speech

---

[3] Why wasn't the poll tax quashed at cabinet? According to King and Crewe, the checks and balances failed: "The policy was gestated and born almost wholly in-house, within one corner of the already secretive and secluded world of Whitehall." It was ultimately waved through during a meeting at Chequers with only half the cabinet in attendance, where few were aware of what was being discussed in advance, and where no papers had been circulated.

was swift, not least among human rights campaigners worried about an extension of police powers. But what few campaigners or, indeed, journalists picked up on was a more prosaic problem, which meant the policy was flawed *on its own terms*. Why? Because a large proportion of thugs would not have a valid debit card, nor anything like as much as £100 in their account. As King and Crewe put it, "The prime minister was assuming that other people lived lives much like his own. His assumption was unfounded."[4]

III

Homophily is pervasive. Our social networks are full of people with similar experiences, views, and beliefs. Even when groups start out with diversity, it can be squeezed out by a process of social osmosis as people converge on the dominant assumptions, a phenomenon known as *assimilation*. The author Shane Snow has shared a telling quote from a senior executive at a major bank:

> She told me, shaking her head, how painful it was to see the company hire all these great college kids—all sorts of backgrounds; all sorts of ideas brimming in their heads—only to watch them gradually remolded to "fit" the culture of the organization. They came with unique insights and voices. She heard those voices fade, unless it was to echo the company's "accepted" way of thinking.

---

[4] It is, of course, possible to come up with historical examples in which a narrow demographic (say, aristocrats or peasants) have come up with enlightened policies, but it would be a mistake to infer that narrow demographics make for better decision-making groups. The problem is that we don't see the counterfactual: Would a more diverse group have made a better decision? This is why diversity science is so important. Randomized trials show that diverse teams systematically come up with superior judgments, better predictions, and wiser strategies.

The clustering of people in small parts of the problem space, then, is a predictable consequence of human psychology. Groups have a built-in tendency to become clone-like. In this sense, the CIA and the poll tax review team are not outliers, they are symptoms. Indeed, look at many cabinets, law firms, army leadership teams, senior civil servants, and even executives at some tech companies. To say that many of these groups are homogenous is not to criticize any individual, it is to note that when smart individuals have overlapping frames of reference, they become collectively myopic.

Wise groups express a different dynamic. They are *not* clone-like. They do not parrot the same views. Instead, they are more like groups of rebels. They do not disagree for the sake of it, but bring insights from different regions of the problem space.

Such groups contain people with perspectives that challenge, augment, diverge, and cross-pollinate. This represents the hallmark of collective intelligence: how wholes become more than the sum of their parts.

In figure 3, the individuals are no smarter than those in the team of figure 2 in the opening section of this chapter. And yet they possess vastly higher levels of collective intelligence. They have *coverage*. And they reveal why, when it comes to complex problems, it is important to work with people who think differently.

The first step for any group seeking to tackle a tough challenge, then, is *not* to learn more about the problem itself. It is not to probe deeper into its various dimensions. Rather, it is to take a step back and ask: Where are the gaps in our collective understanding? Are we beset by conceptual blinkers? Has homophily pulled us into one tight corner of the problem space?

Unless this deeper question is confronted, organizations run the risk of a pervasive glitch in group deliberation: examining a problem, looking ever deeper, while doing little more than reinforcing their blind spots.

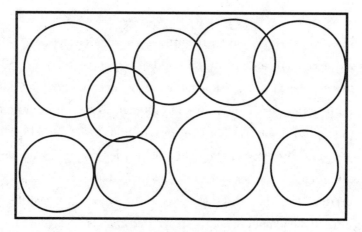

Figure 3. An intelligent team
(A team of rebels)

We need to address cognitive diversity *before* tackling our toughest challenges. It is only then that team deliberation can lead not to mirroring, but to enlightenment.

KARLSKOGA IS A beautiful town in northern Sweden, on the northern edge of Lake Möckeln. It is replete with woodland and fine buildings. I traveled there often when I lived in Sweden in my late teens, and found the place captivating.

Anybody who has spent time in Sweden will know that one of the most important local government policies is snow clearing. On average, Stockholm, the capital, receives around 170 days of precipitation, with much of it occurring in autumn and winter. I remember spending many mornings helping my roommates shovel snow from the driveway. For decades, the snow-clearing operation in Karlskoga followed what seemed like a logical approach. It started with the major traffic arteries and ended with pedestrian walkways and bicycle paths. The (mostly male) officials in the town council wanted to make the daily commute as efficient as possible. They were looking out for the interests of their electorate.

But the council came to an unusual but perceptive realization. They were too homogenous. Remember that in policymaking, which affects huge numbers of people, demographic differences help inform deliberations. In her fine book *Invisible Women*, Caroline Criado Perez highlights that when more women were brought into decision-making positions, collective intelligence started to undergo a remarkable shift.

A fresh analysis revealed that the sexes, on average, travel differently, something that had not previously struck the officials. Men typically take the car to work while women are more likely to take public transport or walk. In France, for example, 66 percent of public transport passengers are women, while in Philadelphia and Chicago it is 64 percent and 62 percent, respectively.

Men and women also have different travel patterns. Men tend to have a twice-daily commute in and out of town in their cars. Women, who do 75 percent of the world's unpaid care work, tend to "drop children off at school before going to work; take an elderly relative to the doctor and do the grocery shopping on the way home," Perez writes. This is sometimes called *trip chaining*. This disparity is found across Europe and is particularly pronounced in households with young children.

As new perspectives were aired, other statistics, previously overlooked, loomed into view. This is important because smart judgments hinge not just on how we interpret data, but on the data we look for in the first place. Statistics from northern Sweden showed that hospital admissions for injuries are dominated by pedestrians, who are hurt three times more often than motorists in slippery or icy conditions. These exert a cost in healthcare and lost productivity. According to one estimate, this accounted for 36 million kronor in Skåne County alone over the course of a single winter. *This is about twice the cost of winter road maintenance.*

With the conceptual blinkers off, Karlskoga took the decision to change decades of policy, prioritizing pedestrians and public transport users on the grounds that "driving a car through three inches of snow is easier than pushing a stroller (or a wheelchair or a bike) through three inches of snow." This was not just better for women, but for the community—and the balance sheet. "Prioritizing pedestrians on the snow-clearing schedule makes economic sense," Perez writes.

Perhaps the key thing to note is that the original schedule was not written out of malice toward women. It wasn't consciously prioritizing drivers above those pushing strollers. No, the problem was perspective blindness. As Perez puts it, "It came about as a result of . . . a gap in perspective. The men . . . who originally devised the schedule knew how they traveled, and they designed around their needs. They didn't deliberately set out to exclude women. They just didn't think about them."

IV

Perhaps the most vivid way to highlight the difference between teams of rebels and teams of clones is through the science of prediction. This may sound like an arcane topic, but predictions are baked into everyday life. Any time an organization decides to do X rather than Y, they are implicitly predicting that X will be better. Prediction is central to pretty much every decision we make, whether at work or in everyday life.[5]

Perhaps the most brilliant study of prediction in recent times was led by Jack Soll, a psychologist at Duke University. He and his colleagues analyzed 28,000 forecasts by professional economists. Their first finding was not at all surprising. Some economists are better performers than others. Indeed, the top forecaster was around 5 percent more accurate than an average forecaster.

---

[5] Predictions also create rich data sets and are wonderfully amenable to mathematical analysis.

But then Soll added a twist. Instead of looking at individual predictions, he took the average prediction of the top six economists. To stretch language a little, these forecasters were being placed into a team. The average of their predictions is what you might call their collective judgment. Soll then checked if this prediction was more accurate than that of the top-ranked economist.

Now, in a simple task, the answer must be no. In a running race, the average time of six runners has to be slower than the time of the fastest runner. This is what Justice Scalia had in mind when he argued for a trade-off between diversity and excellence. But this analysis flips when we move from simple to complex problems. Indeed, when Soll compared the collective judgment of six economists with the judgment of the top economist, it was not less accurate, it was more accurate. And not just a little more accurate but *15 percent more accurate.* This is a staggering difference—so large, in fact, that it shocked the researchers.

Think back to the experiment in chapter 1 when Japanese and American people look at underwater scenes. You'll remember that they tended to see different things. Why? Because Americans and Japanese, on average, have different frames of reference. This is why combining these different perspectives creates a more comprehensive picture.

It turns out that economic forecasters have frames of reference, too. These are sometimes called *models.* A model is a way of making sense of the world: a perspective, a point of view, often expressed as a set of equations. No economic model is complete, however. Each model contains blind spots. The economy is complex (unlike, say, the orbit of Jupiter, which can be precisely predicted). The rate of industrial production, for example, hinges on the decisions of thousands of businesspeople, operating tens of thousands of factories and firms, and influenced by millions of variables. No model can account for all this complexity. No economist is omniscient.

But this implies that if we bring different models together, we create a more complete picture. No economist has the whole truth, but a group of diverse economists gets closer to the truth. Often, much closer. With prediction tasks, this is known as the *wisdom of crowds*. There are now dozens of examples of this aspect of diversity science. When, for example, the researcher Scott Page asked his students to estimate the length in miles of the London Underground by writing their guesses on slips of paper, the collective prediction was 249 miles. The true value is 250 miles.

Group wisdom emerges whenever information is dispersed among different minds. Think of the students guessing the length of the London Underground: one may have visited London, another may have familiarity with the subway in New York, and so on. When people make estimates, they are translating whatever information they have into a number. Each guess adds to the pool of information.

Of course, each person is also contributing mistakes, myths, and blind spots. This creates a pool of error almost as big as the pool of information. But the information is, by definition, pointing toward the correct answer. The errors, on the other hand, emerge from different sources and point in different directions. Some estimates are too high, others are too low, tending to cancel each other out. Again, as Philip Tetlock puts it, "With valid information piling up and errors nullifying themselves, the net result is an astonishingly accurate estimate." James Surowiecki, who has written a fine book on group decision making, puts it this way: "Each person's guess, you might say, has two components: information and error. Subtract the error, and you're left with the information."

Of course, if the individuals in the group don't know much, then combining their judgments won't achieve much. If you ask a group of laypeople to estimate how much ocean levels will rise over the next decade, you won't get very far. To achieve group wisdom, you need wise individuals. But you also need individuals without the same blind spots. To return to economic forecasting: suppose you could identify and clone the most accurate forecaster in the world. If you were putting together a team of six forecasters, would it make sense to put six of these clones together? On the surface, the team sounds unbeatable. Each member is more accurate than any forecaster in any other team. Isn't this the perfect team?

We can now see that the answer is an emphatic no! They all think in the same way. They use the same model and make the same mistakes. Their frames of reference overlap. Indeed, the Soll experiment implies that a diverse group of six forecasters, while individually less impressive, would be 15 percent more accurate.

It is worth pausing to reflect on how potentially world-changing this result is. It reveals—in precise, mathematical terms—the sheer power of cognitive diversity. A team of world-class forecasters who think in the same way are less dramatically intelligent than a group of forecasters who think differently.

Of course, most of us do not sit around the table at work, or in life, making numerical forecasts of the kind familiar to economists. But we do try to solve problems, come up with creative ideas, determine strategies, spot opportunities, and much else besides. This is the essence of the group-based work that is coming to dominate our world. And yet we can expect diversity to have even *stronger* effects on these tasks.

Let us take creativity and innovation. Ask yourself this question: Suppose that you put together a team of ten people to come up with ideas to solve, say, the obesity crisis. Suppose, too, that each of these ten people comes up with ten useful ideas. How many useful ideas do you have in total?

In fact, this is a trick question. You can't infer the number of ideas in a group from the number of ideas of its members. If these people are clone-like and come up with the same ten ideas, you have only ten ideas overall. But if the ten people are diverse and come up with different ideas, you could have one hundred useful ideas. That is not 50 percent more ideas, or 100 percent more ideas, but almost 1,000 percent more ideas. This is another huge effect *solely attributable to diversity.*

In problem-solving teams, we see the same pattern. We noted that in prediction tasks, taking the average of independent forecasts is an effective way of pooling information. With problem-solving, however, averaging is often a terrible idea. Taking the average of two proposed solutions can often lead to incoherence. This is the origin of the phrase "a camel is a horse designed by committee." With most problems, a team has to reject some solutions in favor of others.

But this again reveals why diversity matters. With homogenous groups people tend to get stuck in the same place. Diverse teams, on the other hand, come up with fresh insights, helping them become unstuck. Rebel ideas are effectively firing the collective imagination. As the leading psychologist Charlan Nemeth puts it, "Minority viewpoints are important, not because they tend to prevail but because they stimulate divergent attention and thought. As a result, even when they are wrong they contribute to the detection of novel solutions that, on balance, are qualitatively better."

But the power of diversity is subtler than even these examples might suggest. The deepest problem of homogeneity is not the data that clonelike teams fail to understand, the answers they get wrong, the opportunities they don't fully exploit. No, it is the questions they are not even asking, the data they haven't thought to look for, the opportunities they haven't realized are out there.

The more challenging the domain, the less that any single person—or perspective—can hope to grasp. With prediction teams, homogenous minds make the same errors. With problem-solving teams, they get stuck in the same place. With strategy teams, they miss the same opportunities.

When Justice Scalia argued that there was a trade-off between performance and diversity, he was making a seductive conceptual error. It is the same error that leads most people to express surprise when told that the average of six forecasters is more accurate than the top forecaster, and that deludes people into thinking that a group of wise individuals must constitute a wise group. Scalia was, in effect, looking at the problem from the individualistic perspective, not the holistic perspective. He didn't take account of the fact that collective intelligence emerges not just from the knowledge of individuals, but also from the differences between them. Let us call this the *clone fallacy.*

The tragedy is that this fallacy is pervasive. Indeed, perhaps the most striking conversation I had while researching this book was with a renowned economic forecaster. I asked if he preferred to work with people who think in the same way, or who think differently. He replied, "If I truly think my model is the best one out there, then I should work with people who think like me."

V

Most organizations have an avowed policy of meritocratic hiring. The idea is to recruit on the basis of skill and potential, rather than on arbitrary factors like social connections, race, or gender. This is both morally commendable and self-interested. Institutions are hiring talent regardless of what it looks like. But it also contains latent dangers. Let us take a hypothetical example to flesh out the logic. Suppose that some universities have a strong reputation for, say, software development. These universities are likely to attract the smartest software students. These students, in turn, will graduate with the most impressive credentials. Now, suppose you are running a top software company. Wouldn't you want these students? Wouldn't you want to pack your organization with the brightest and the best?

The enlightened answer, once again, is no. These graduates will have studied under the same professors and absorbed similar insights, ideas, heuristics, and models—and perhaps worldviews, too. This is sometimes called *knowledge clustering.* By selecting graduates in a meritocratic way, organizations can find themselves gravitating toward clone-like teams. This is not to dismiss meritocracy. It is merely to point out that collective intelligence requires both ability *and* diversity.

Indeed, no test that ranks individuals can—on its own—construct intelligent groups, another point made by Scott Page: "Suppose you are building a team to come up with creative ideas. First, any test applied to an individual can only measure that individual's ideas. Second, a clone of the person who scores highest on whatever test we apply necessarily adds less to the group than a second person with a single different idea. Therefore, no test can exist."

Now let us return to the distinction between cognitive diversity (differences in thoughts, insights, perspectives) and demographic diversity (differences in race, gender, class, and so on). We noted in chapter 1 that demographic diversity often overlaps with cognitive diversity. This is intuitive since our identities influence our experiences, perspectives, and more. Advertising firms, for example, rely on demographic diversity to create campaigns that appeal to the breadth of their client base.

This helps explain the previously mentioned study by Chad Sparber (along with dozens of others), which found that an increase in racial diversity of just one standard deviation increased productivity by more than 25 percent in legal services, health services, and finance. In any domain that requires an understanding of broad groups of people, demographic diversity is likely to prove vital.

But there are other contexts in which the overlap is less significant, or even nonexistent. In the very same piece of research, Sparber found that increases in racial diversity offered no efficiency gains for firms producing aircraft parts, machinery, and the like. Why? Because the experience of being, say, Black provides few, if any, novel insights into the design of, say, engine parts.

We can make this point in a different way with economic forecasting. Take two economists: one white, gay, male, and middle-aged, the other Black, young, female, heterosexual. These economists are different in demographic terms—and might tick all the boxes on a conventional diversity matrix. But suppose they went to the same university, studied under the same professor, and left with similar models. In these circumstances, they would be clone-like in relation to the problem.

Now take two white, middle-aged, bespectacled economists, who have the same number of children and like the same TV programs. They may seem homogenous and, from a demographic perspective, they are. But suppose that one is a monetarist and the other a Keynesian. These are two different ways of making sense of the economy, two very different models. Their collective prediction will, over time, be significantly better than either alone. The two economists may look the same, but they are diverse in the way they think about the problem.

This is worth keeping in mind because hiring someone who is different in terms of color or gender does not guarantee an increase in cognitive diversity. Building collective intelligence cannot be reduced to a box-ticking exercise. Consider, too, that people who start out diverse can gravitate toward the dominant assumptions of the group. This can lead to a situation where leadership teams look diverse, but are—in cognitive terms—anything but. The team members have all been at the organization so long that they have come to share nearly identical views, insights, and patterns of thinking.

Successful teams are diverse, but not *arbitrarily* diverse (see fig. 4). A group of scientists designing a hadron collider is unlikely to benefit from

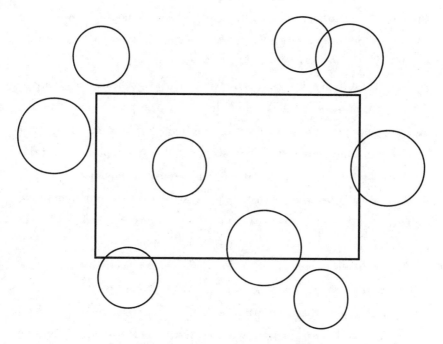

Figure 4. A diverse but collectively unintelligent team

hiring, say, a skateboarder, of whatever color or gender. The skater would have diverse information, but this would not likely impinge on the problem space.

Diversity contributes to collective intelligence, then, but only when it is relevant. The key is to find people with perspectives that are both *germane* and *synergistic.*

For economic forecasters, collective intelligence emerges from accurate predictors who deploy different models. For an intelligence agency, it emerges from outstanding analysts who possess a rich diversity of experience, the better to understand the multiplicity of threats they face. For policymakers, it emerges from exceptional individual politicians with (among other things) backgrounds that span the demographic spectrum of the electorate they serve. For teams working in other contexts—well, we will see further examples as the book progresses.

The most important point is the *generalized* significance of diversity. Diversity isn't some optional add-on. It isn't the icing on the cake. Rather, it is the basic ingredient of collective intelligence. You can see the power of diversity from a broader perspective, too. Diversity explains why price systems work so effectively and why open-source innovation platforms and wikis have become pervasive. These all share the same underlying signature: they aggregate the disparate information contained in different minds.[6]

Diversity has even moved to the heart of artificial intelligence. A couple of decades ago, machine learning was based on single algorithms. Today, it is largely characterized by ensembles of diverse predictors. Scott Page hit upon the same pattern when creating problem-solving computer models. "I stumbled on a counterintuitive finding," he said. "Diverse groups of problem-solvers . . . consistently outperformed groups of the best and brightest."

VI

One of the much-vaunted solutions to a lack of diversity in politics and beyond is the use of focus groups. These are often hailed as a means of offering the benefits of diversity without having to dilute the clubby homogeneity of elite power structures. The basic idea is that you put a representative cohort of people in a room, ask questions, find out what they like

---

[6] This is a point that has been made by the economist Friedrich Hayek, who showed how prices emerge as a consequence of independent judgments by lots of different people acting on their own information and preferences. Often, market prices do an astonishingly effective job at combining diffuse information.

and what they don't, note any objections and practical problems, and then refine the policy accordingly. Advertisers sometimes use this approach with market research, testing the ideas on a diverse audience to gain insights into what works and what doesn't.

But it should be clear that such approaches, while sensible in their own terms, miss the deeper point. Why? Because diversity is not just about getting answers from focus groups or market research. It is about the questions that are asked in the first place, the data that is used as the basis for deliberation, and the assumptions that permeate our approach to any issue.

This is true of not just politics, but science—supposedly the most objective of disciplines. One survey of sports science journals found that 27 percent of studies focused exclusively on men, and only 4 percent on women. It is no coincidence that the vast majority of sports scientists are men. This is one tiny example of how biases can be baked into deliberations before scientists start to answer questions, and where data is skewed before the lessons are probed. This shows that while demographic diversity and cognitive diversity are conceptually distinct, they typically overlap.

You can see the same point in a different way by looking at primatology. Before Jane Goodall came on the scene, the field was dominated by men. They adopted Charles Darwin's view of evolution, focusing on competition among males for access to females. In this framework, female primates are passive, and the alpha male has access to all the females, or females simply choose the most powerful male. But this frame of reference contained a blind spot. Only when a critical mass of women arrived on the scene did primatology come to realize that female primates are far more active and might even have sex with many males, insights that created a richer, more explanatory theory of primate behavior.

Why did female scientists see something that men had missed? In her fascinating book *The Woman That Never Evolved,* the anthropologist Sarah Blaffer Hrdy writes, "When, say, a female lemur or bonobo dominated a male, or a female langur left her group to solicit strange males, a woman fieldworker might be more likely to follow, watch, and wonder than to dismiss such behavior as a fluke."

We saw in the opening chapter that Japanese people tend to focus more on context and less on individuals when compared with Americans. It is noteworthy that primatology has benefited from this very effect. As the academics Douglas Medin, Carol D. Lee, and Megan Bang put it in a lead article for *Scientific American*:

In the 1930s and 1940s U.S. primatologists . . . tended to focus on male dominance and the associated mating access. Rarely were individuals or groups tracked for many years. Japanese researchers, in contrast, gave much more attention to status and social relationships, values that hold a higher relative importance in Japanese society. This difference in orientation led to striking differences in insight. Japanese primatologists discovered that male rank was only one factor determining social relationships and group composition. They found that females had a rank order, too, and that the core of the group was made up of lineages of related females, not males.

This takes us back to something mentioned earlier. Remember the warning of John Cleese? "Everybody has theories," he said. "The dangerous people are those who are not aware of their own theories. That is, the theories on which they operate are largely unconscious." We can now see that this applies as much to science as to anything else. In his subtle and beautiful book *Conjectures and Refutations,* Karl Popper, perhaps the greatest philosopher of science, makes this point. His words are among my favorite ever written, and a useful jolt to not just scientists, but all of us:

> Twenty-five years ago I tried to bring home the point to a group of physics students in Vienna by beginning a lecture with the following instructions. "Take pencil and paper; carefully observe, and write down what you have observed!" They asked, of course, what I wanted them to observe. Clearly the instruction, "Observe!" is absurd. . . . Observation is always selective. It needs a chosen object, a definite task, an interest, a point of view, a problem. . . . For a scientist [a point of view] is provided by his theoretical interests, the special problem under investigation, his conjectures and anticipations, and *the theories which he accepts as a kind of background: his frame of reference, his "horizon of expectation"* [my italics].

VII

Denniston was known to his colleagues as A.G.D. A diminutive Scot with a keen mind, he was the man selected to head up the recruitment operation for Bletchley Park. Perhaps his deepest insight, and one that would have seismic implications, was that solving a complex, multidimensional problem requires cognitive diversity.

Instead of looking only for a group of Alan Turings—if such a group existed—Denniston cast his net wider than many thought sensible or desirable. As Michael Smith notes in *The Secrets of Station X*, his recruits included Leonard Foster, a scholar of German and the Renaissance, Norman Brooke Jopson, a professor of comparative philology, Hugh Last, the historian, and A. H. Campbell, a legal philosopher. He also tapped up J. R. R. Tolkien, a professor of Anglo-Saxon at Oxford. (Although Tolkien took an instructional course at the London headquarters of the Government

Code and Cypher School, he ultimately decided to stay in Oxford. Cryptography's loss was literature's gain: during the war years Tolkien would write the bulk of *The Lord of the Rings*.)

The Bletchley Park team was diverse across multiple dimensions. They had different intellectual backgrounds, but also demographic backgrounds. Turing was gay at a time when homosexuality was illegal. The majority of the staff were women, albeit often in administrative roles (Bletchley Park was by no means immune from the sexism in broader society). There were many high-ranking Jewish cryptanalysts. There were also people of other religions and social backgrounds.

Why did any of this matter when it came to cracking a code? Wasn't it just about logic and number crunching? In fact, like all complex tasks, the challenge hinged on multiple layers of insight. Take the conundrum that became known as Cillies. These were sequences of three letters used by German signal operators for the message settings on the machines, for which they would often use girlfriends' names, or perhaps the first three letters of a swear word. They were called Cillies because one of the first spotted was CIL, an abbreviation of Cillie, a German girl's name. These "tells" would help the team narrow down the task of cracking the code.

Code breaking is not just about understanding data, it is also about understanding people. "One was thinking all the time of the psychology of what it was like to be in the middle of the fighting when you were supposed to be encoding a message for your general and you had to put three or four letters in these little windows, and in the heat of the battle you would put up your girlfriend's name or dirty four-letter German words," one of the young female code breakers said. "I am the world's expert on dirty German four-letter words!"

The desire to recruit crossword enthusiasts emerged from the same quest to gain insights from across the problem space. It might have seemed almost frivolous that the recruiters at Bletchley Park were scrutinizing the *Daily Telegraph* crossword competition in a time of war. But they were operating from the holistic perspective. They had made the vital leap of imagination that crosswords have critical features in common with cryptography.

"Whether it is a simple code, or something as complex as the Enigma cipher, which the Bletchley code breakers were working on, the trick is making links between letters and words," Smith said. "Crosswords are the same sort of lateral-thinking exercise."

Shortly before her death in 2013, Mavis Batey, who helped crack the Italian Enigma, which would prove crucial to Britain winning the Battle of Cape Matapan, gave an interview that showcased her capacity for lateral thinking. "My daughter worked in the Bodleian Library," she said, "and one day, she mentioned she had been working on J floor. 'J,' I said, 'ten floors down.' And she looked at me oddly and asked how I could have instantly known that."

There was a human element, too, a point made by the science writer Tom Chivers:

Crosswords are about getting inside the mind of your opponent, and in the same way, code breaking was about getting inside the mind of your enemy. The code breakers came to know the people encoding the messages individually, by their styles, as crossword solvers come to know setters. Mavis Batey worked out that two of the Enigma machine operators had girlfriends called Rosa.

The letter that arrived on Sedgewick's doorstep wasn't a punt. It wasn't about diversity for the sake of it. This was diversity precision-engineered to maximize collective intelligence. "It took imagination to bring together the different minds to solve a fiendishly difficult problem," Michael Smith, who was an intelligence officer before becoming a journalist and author, told me.

To put it another way, cracking the Enigma code relied on cracking a prior code: the diversity code. How easy it would have been to hire only brilliant individuals of a similar ilk. How easy to hire mathematicians who were superb at analyzing data from the Enigma machines, but might not have stopped to wonder about their human operators. By taking a step back, by pondering the blind spots in any perspective, by having the ingenuity to seek insights across the universe of useful ideas, Bletchley Park came to express a collective intelligence of an unusual and remarkable kind.

George Steiner, the philosopher and critic, described Bletchley Park as "the single greatest achievement of Britain during 1939–45, perhaps during the twentieth century as a whole." Bill Bundy, an American code breaker who worked at Bletchley and would go on to become assistant secretary of state in the U.S. government, said that he had never worked with a group of people that was "more thoroughly dedicated and with such a range of skills, insight, and imagination."

After he received his letter, Sedgewick took up the invitation to visit Colonel Nicholls of the general staff, who also happened to be the head of MI8, the British military intelligence department. "I arranged to attend at Devonshire House in Piccadilly, the headquarters of MI8, and found myself among a few others who had been contacted in the same circumstances," Sedgewick later said. "Thus it was that I reported to 'the spy school' at 1 Albany Road, Bedford."

Once he arrived at Bletchley Park, Sedgewick was put to work in Hut Ten, which focused on intercepting weather codes. These were crucial for Bomber Command of the Royal Air Force, helping them make more informed operational decisions, but they had an additional purpose. They were used as cribs for the Enigma machine used by the German Navy.[7]

Cracking this code turned out to be of incalculable significance, playing a key role in the Battle of the Atlantic. It enabled the convoys from America to elude the German U-boats that lay in wait, creating an umbilical link between the United States and Europe, thus enabling Britain to benefit from the merchant supplies that were crucial to continue fighting. One source estimates that it saved up to 750,000 tons of shipping in December 1942 and January 1943 alone.

"When I talked to Sedgewick a few years before his death, what struck me the most was his modesty and sense of duty," Michael Smith told me. "He had a rather mundane job before the war, so being recruited to Bletchley Park represented a fascinating challenge. My impression is that he had the time of his life, working with a remarkable team on the most important of missions."

And this is how a quietly spoken clerk who learned to solve crosswords on his daily commute helped defeat Nazi Germany. Stanley Sedgewick was a member of one of history's finest teams of rebels.

---

[7] The security of the naval Enigma had been strengthened in early 1942, making it particularly difficult to crack.

CONSTRUCTIVE
DISSENT

I

Shortly after midnight on May 10, 1996, Rob Hall and his team entered
the Death Zone. From the South Col, the pitiless expanse of rock-hard
ice and windswept boulders where they had camped overnight in gale-
force winds, it was 3,117 vertical feet to the summit of the world's highest
mountain. If everything went according to plan, they would be stepping
onto the apex of Mount Everest, with its Buddhist prayer flags and as-
sorted mementos, in twelve hours' time.

In addition to Hall, a bearded thirty-five-year-old who was leader of
the expedition, there were two further guides (Andy Harris and Mike

Groom), Sherpas, and eight clients. The clients were experienced climbers, but didn't have the world-class technical credentials to climb Everest
unaided. Their number included Jon Krakauer, an author and adventurer
who was writing up the expedition for *Outside* magazine, Beck Weathers,
a pathologist from Texas, with ten years of mountaineering experience,
and Yasuko Namba, a forty-seven-year-old businesswoman from Tokyo
who had climbed the tallest mountains on six of the world's seven conti-
nents. A successful ascent of Everest would take her into the record books
as the oldest woman to complete all seven summits.

Hall was confident in his team and his preparations. He had summited
Everest four times previously and combined supreme technical skill with
agility and strength. He had met Jan, his wife, on the way to an Everest
attempt in 1990 (a doctor, she was working at a clinic below base camp)
and fell in love. "I asked Jan to go out with me as soon as I got down
from Everest," he would later say. Their first date was climbing Denali in
Alaska, and they married two years later. In 1993, they summited Everest
together, only the third married couple to do so.

Jan usually worked out of base camp during Hall's attempts on Everest, but this time she had to turn down the chance. She was seven months
pregnant. That gave the climb an even greater sense of anticipation for
Hall. When he returned home to New Zealand, he would experience the
thrill of becoming a father for the first time. "I can't wait," he said.

But Hall was experienced enough to know that every step upward
would take the team further into peril. The South Col is five miles above
sea level, the air so thin that the climbers were now using bottled oxygen,
masks strapped to their faces as their bodies cowered against the ferocious
demands of the troposphere. "Every minute you remain at this altitude
and above, your minds and bodies are deteriorating," Hall told his team.
Krakauer would write, "Brain cells were dying. Our blood was growing

dangerously thick and sludge-like. Capillaries in our retinas were spontaneously hemorrhaging. Even at rest, our hearts beat at a furious rate."

As the group looked up at the famous triangular face of the peak known locally as Chomolungma, "Goddess Mother of the World," they knew the technical challenges on the final ascent would be exacting. First, the patient climb to the Balcony, headlamps illuminating the route, ropes traversing the slope, the potential for a deadly rockfall an ever-present fear. Then, the ascent to the South Summit, steep and continuous, the rising sun bathing Lhotse to the south in phantasmagorical light.

And then, just beneath the summit proper, the Hillary Step. Named after Sir Edmund Hillary, the first person to make the top alongside Sherpa Tenzing, it is the most famous vertical face in all climbing. "The most formidable-looking problem on the ridge—a rock step some forty feet high," Hillary would write. "The rock itself, smooth and almost holdless, might have been an interesting Sunday afternoon problem to a group of expert climbers in the Lake District, but here it was a barrier beyond our feeble strength to overcome."

Everest is not, according to insiders, the most beautiful of the world's peaks. It is bulky and prosaic compared with the soaring silhouettes that compose many of the great mountains. But what it lacks in aesthetic appeal, it makes up for in mystique. Krakauer writes, "I stared at the peak for perhaps thirty minutes, trying to apprehend what it would be like to be standing on that gale-swept vertex," as he beheld the massif on the trek from the Lukla airstrip to base camp. "Although I'd ascended hundreds of mountains, Everest was so different from anything I'd previously climbed that my powers of imagination were insufficient for the task. The summit looked so cold, so impossibly far away."

Everest is also deadly. Since the mountain had first been attempted by a British expedition in 1921, 130 climbers had died, an attrition rate of one fatality for every four climbers to reach the top. Perhaps the most famous death was also one of the earliest, that of George Mallory[8] in 1924. With rudimentary equipment but astonishing courage, the Englishman dared the final ascent on June 8 along with his companion Andrew "Sandy" Irvine.

The top of the mountain that day was wreathed in mist, preventing the supporting team from observing their progress, but at 12:50 p.m. the clouds parted for a few moments. Noel Odell, one of their teammates, witnessed them high up on the northeast ridge, five hours behind schedule but "moving deliberately and expeditiously" toward the peak. Neither Mallory nor Irvine was seen again until 1999, when Mallory's corpse was found at 26,760 feet on the north face. The consensus of historians is that neither man made it to the summit.

The dangers were palpable to every member of Hall's team. They had witnessed the dead bodies that litter the mountainside, and had received stern warnings about the importance of supplementary oxygen. Since arriving at base camp at 17,600 feet, they had made three acclimatization climbs. The first up the Khumbu Icefall—full of crevasses, moving ice, and the threat of avalanche—had taken them to 19,500 feet. The second and third had taken them first to 21,000 feet, then 23,500, each hour spent at altitude forcing their bodies and minds to become more accustomed to air that, at the peak itself, would contain a mere third of the oxygen at sea level.

But now, well into the so-called Death Zone above 26,000 feet, they were in the most forbidding territory in all mountaineering. Hall had already decided that the team would require a turnaround time of 1 p.m., or 2 p.m. at the latest. If they hadn't made it to the top by then, they would have to head back down. This wasn't a technical judgment so much as a

---

[8] It was Mallory who responded to a journalist badgering him for the reason he wished to risk life and limb to climb Everest with the immortal line: "Because it's there!"

mathematical one. With three oxygen canisters per person, each providing around six to seven hours of gas, any later would be flirting with calamity. As Hall put it, "With enough determination, any bloody idiot can get up this hill. The trick is to get back down alive."

The other complicating factor that day was that other teams were also attempting the summit, a common occurrence due to the global fascination of the Himalayan peak. The Mountain Madness team was led by Scott Fischer, an amiable American and one of the most skilled alpine mountaineers in the world. His guides were also superb at their trade, including Anatoli Boukreev, who had climbed Everest twice. Among the clients were Sandy Pitman, an American mountaineer who, like Namba, had completed six of the seven summits. She was doing a daily video blog for NBC. Also on the slopes that day was a team, albeit much smaller, from Taiwan.

By the time the sun peaked over the rim of the horizon at 5:15 a.m., Krakauer, a member of Hall's group, had reached the crest of the southeast ridge. "Three of the world's five highest peaks stood out in craggy relief against the pastel dawn," he would later write. "My altimeter read 27,600 feet." It was a glorious sight, but elsewhere on the slopes tiny problems were beginning to accumulate.

Ropes hadn't been preinstalled above 27,400 feet, which led to logjams while they were put in place. Boukreev, Neal Beidleman (a guide in the Mountain Madness team), and Sherpas painstakingly payed out the rope on the exposed upper sections. Meanwhile, Fischer was farther down the mountainside. He had expended much energy three days earlier helping his friend Dale Kruse, who had become ill, down to base camp. He was also exhibiting symptoms consistent with high-altitude pulmonary edema, a buildup of fluid in the lungs.

It wasn't until a little after 1 p.m. that Krakauer, ahead of the rest of his team, made it to the summit. He was thrilled to have fulfilled a lifetime ambition, but he could sense that the moving parts of the expedition were becoming misaligned. Hall was still well below the summit. Pitman and other team members were becoming ever more tired. The deadline for a safe turnaround time was rapidly approaching. Wispy clouds were filling the valleys to the south.

And yet, perhaps even then, none of the climbers could have guessed that in the coming hours, eight of their number would lose their lives in one of the most infamous days in the history of the world's most famous mountain: the 1996 Everest disaster.

IN THE YEARS SINCE 1996, many survivors have told their accounts. Krakauer wrote the bestseller *Into Thin Air.* Weathers wrote *Left for Dead.* IMAX made a documentary called *Everest,* while National Geographic made a feature called *The Dark Side of Everest.* In 2015, the disaster was made into a Hollywood blockbuster starring Jason Clarke, Josh Brolin, and Keira Knightley. *Everest* grossed more than $200 million at the box office.

And yet despite the plethora of accounts, there is, to this day, no consensus about what went wrong and what, by implication, should be learned. Krakauer was deeply critical of Boukreev, a guide in the Mountain Madness team, who had advanced too far ahead of his clients. Boukreev hit back with his own book entitled *The Climb* and was defended by many of the most authoritative voices in mountaineering. Pitman, who spent years haunted by what had happened, complained that various accounts had assassinated her character. Krakauer, for his part, said that his depiction in *Everest* (he was played by the actor Michael Kelly) was "total bull."

Differences of this kind are, perhaps, inevitable, particularly when there is a desire to apportion blame. People had died, families were bereaved, and many were confused as to how things had gone so badly. It is common for first-person accounts to diverge in the aftermath of a disaster, sometimes profoundly. But in this chapter, we are going to look at the possibility that all of these accounts are wrong. We will examine the idea that the problem wasn't with the actions of any individual, but in the way they communicated.

In the opening two chapters, we examined how different perspectives can enlarge collective intelligence, often in idiosyncratic ways. Sometimes, however, the benefits of diversity are more prosaic. On a mountainside, different climbers are at different positions on the slopes and are seeing different things. A climber at one point will observe the energy levels of nearby climbers, problems in the vicinity, clouds rolling in from the west. These will not be visible to those at different points of the mountain. One person has a single pair of eyeballs. A team has many. So we are going to ask how useful information and perspectives are *combined*. For diversity to work its magic, different perspectives and judgments must be expressed. Useful information that never gets aired is not useful.

There is also the question of who makes the final decision once the various perspectives have been expressed. If there are competing views, whose wins out? If there are different insights, do we fuse them together, or select one rather than the other? In this chapter, we will move from the conceptual foundations of diversity to the practical implementation.

Everest is an apt setting for this purpose. Weather conditions are inherently uncertain. No matter how much planning and preparation you have done, there are unexpected twists and turns. The moving parts as conditions morph make huge demands not merely on physical endurance, but cognitive load. Mountaineering is, in this sense, what theorists call a VUCA environment: volatile, uncertain, complex, and ambiguous.

II

Psychologists and anthropologists don't agree about much, but one thing they do agree on is the significance of hierarchies. Humans share dominance hierarchies with other primates and, according to the psychologist Jordan Peterson, even lobsters. "The presence of hierarchy stretches back across tens of thousands of generations to the advent of *Homo sapiens* and, indeed, much further to include other primate species," Jon Maner, professor of psychology at Florida State University, has said. "The human mind is, quite literally, designed to live within hierarchically arranged groups."

The emotions and behaviors associated with dominance hierarchies are so deeply written into our minds that we scarcely notice they are there. Dominant individuals adopt more expansive gestures, issue threats, and motivate subordinates through fear. Particularly dominant alphas raise their voices, gesticulate, and bare their teeth. This is as true of many bosses in the financial district as of an alpha in a chimpanzee troop. Those in lower positions tend to signal subservience with lowered heads, hunched shoulders, and gaze avoidance—what George Orwell termed *cringing*.

Indeed, so highly attuned is our status psychology that you can place five strangers in a room, give them a task, and watch dominance hierarchies developing within seconds. What is even more remarkable is that external observers, who can't even hear what is being said, can accurately place people at the various positions in the hierarchy, just by watching their postures and expressions.

Hierarchy is not just what we do, it is who we are.

The pervasiveness of dominance hierarchies hints that they serve an important evolutionary purpose. When the choices that confront a tribe or group are simple, it makes sense for a leader to make decisions and for everyone else to fall into line. This boosts speed and coordination. Tribes with dominant leaders tended to fare best in our evolutionary history.

But in situations of complexity, dominance dynamics can have darker consequences. As we have seen, collective intelligence hinges on the expression of diverse perspectives and insights—what we have called *rebel ideas*. Diverse expression can shut down in a hierarchy where dissent is perceived by the alpha as a threat to their status. Dominance, in that sense, represents a paradox: humans are inherently hierarchical, and yet the associated behaviors can thwart effective communication.

An incident that brings this paradox to light involves United Airlines 173,[9] a flight that took off from Denver on December 28, 1978, flying to Portland, Oregon. Everything went smoothly until the final approach. The captain pulled the lever to lower the landing gear, but instead of a smooth descent of the wheels, there was a loud bang, and a light that should have indicated the landing gear was down and secure failed to light up. The crew couldn't be sure that the wheels were down, so the captain put the plane in a holding pattern as they attempted to troubleshoot the problem.

They couldn't see below the plane to check if the wheels were down, so they conducted proxy checks. First the engineer went into the cabin. When the landing gear has slid down into place, two bolts shoot up above the wing tips. These bolts were, indeed, up. They then contacted the United Airlines Control Center in San Francisco to talk through what had happened, and received advice that the wheels were probably down.

But the captain still wasn't certain. What had caused that loud bang? Why hadn't the light on the dashboard turned on? Landing without the wheels in place can generally be achieved without loss of life, but it contains risk. The captain, a decent man with extensive experience, didn't want

---

[9] I refer to this incident in my book *Black Box Thinking* in the context of safety investigations.

to place his passengers in unnecessary danger. He began to wonder if the reason the light had failed to turn on was because of the wiring. Or perhaps it was a faulty bulb.

However, as he deliberated and the plane continued its holding pattern, a new danger had come into play. The plane was running out of fuel. The engineer knew that the fuel was critical: he could see it disappearing on the gauge before his eyes. He also had a powerful incentive to alert the pilot: his life, and the lives of everyone else on the plane, were on the line.

But this was the 1970s. The culture of aviation was characterized by a dominance hierarchy. The pilot was called "sir." The other crew members were expected to defer to his judgments and act upon his commands. This is what sociologists call a *steep authority gradient*. If the engineer voiced his concerns about the fuel, it might have carried the implication that the pilot wasn't on top of all the key information (which he wasn't!). It might have been perceived as a threat to his status.

By 17:46 local time, the fuel had dropped to five on the dials. This was now an emergency. Almost two hundred lives, including that of the engineer, were in severe danger. The pilot was still focused on the bulb, oblivious to the dwindling fuel. Perception had narrowed. You might suppose that the engineer might have said: "We have to land now! Fuel is critical!" But he didn't. We know from the cockpit voice recorder that he merely hinted at the problem. "Fifteen minutes is gonna really run us low on fuel, here," he said.

The engineer was so fearful of directly challenging the captain that he softened his language. The captain interpreted his remarks as meaning that while the fuel was going to get low as they circled again, it wasn't going to run out. This was incorrect, and the engineer knew it. Even at 18:01, when it was probably too late, and with the captain now focused on the plane's antiskid system, the engineer and first officer were still strug-gling to clearly state the problem.

It wasn't until 18:06, with the engines flaming out, that they finally made the information explicit, but it was too late. They had gone past the point of no return, not because the team lacked the information, but because it wasn't shared. The plane crashed minutes later, piling into a wooded suburb, ploughing through a house, and coming to rest on another house. The lower left side of the fuselage was completely torn away. On a clear evening, where the airport had been visible since they entered the holding pattern, more than twenty people died, including the engineer.

Now, this may seem like a freak event, but the psychology is univer-sal. According to the National Transportation Safety Board, more than thirty crashes have occurred when copilots have failed to speak up. In one wide-ranging analysis of twenty-six different studies in healthcare, it was found that a failure to speak up was "an important contributing factor in communication errors."

This isn't just about safety-critical industries. It is about the human mind. "People often think their own industries are very different," Rhona Flin, professor of applied psychology at Aberdeen University, has said. "Actually, if you're a psychologist who's worked in different industrial set-tings, it all looks pretty much the same. . . . They're all humans working in these technical environments. They're affected by the same kind of emotions and social factors."

In an experiment conducted not long after the Portland crash, re-searchers observed crews interacting in flight simulators, and the same problem kept re-emerging. "Captains were briefed in advance to take some bad decisions or feign incapacity—to measure how long it would take for copilots to speak up," Flin said. "One psychologist monitoring their re-sponses commented, 'Copilots would rather die than contradict a captain.'"

On the surface, the willingness to risk death rather than challenge the alpha may seem odd. Certainly something that wouldn't affect you, or me. But the failure to speak up can happen unconsciously. We do it automatically. Think of any workplace. Those in subordinate positions seek to please the boss, parroting his thoughts and even hand gestures. This eliminates diverse insights, not because they do not exist, but because they are not expressed.

A clever study by the Rotterdam School of Management analyzed more than three hundred real-world projects dating back to 1972 and found that projects led by junior managers were more likely to succeed than those with a senior person in charge. On the face of it, this seems astonishing. How could a team perform better when deprived of the presence of one of its most knowledgeable members?

The reason is that this leadership comes at a sociological price when linked to a dominance dynamic. The knowledge squandered by the group when the senior manager is taken out of the project is more than compensated for by the additional knowledge expressed by the team in his absence. As Balazs Szatmari, lead author of the study, put it, "The surprising thing in our findings is that high-status project leaders fail more often. I believe that this happens not despite the unconditional support they get, but actually because of it."

The Indian tech entrepreneur Avinash Kaushik has an evocative phrase for the way dominance dynamics influence many organizations. He uses the acronym HiPPO: Highest-Paid Person's Opinion. "HiPPOs rule the world, they overrule your data, they impose their opinions on you and your company customers, they think they know best (sometimes they do), their mere presence in a meeting prevents ideas from coming up," he said.

We can see a dominance dynamic in figure 5. This is an impressively diverse team; they have plenty of coverage across the problem space. And yet when brought under a dominant leader (the dark circle), subordinates

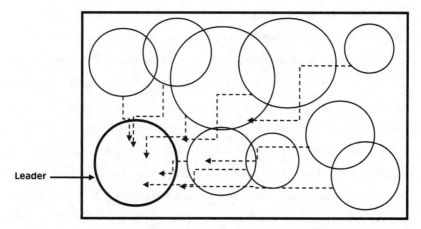

Figure 5. A diverse team with a dominance dynamic

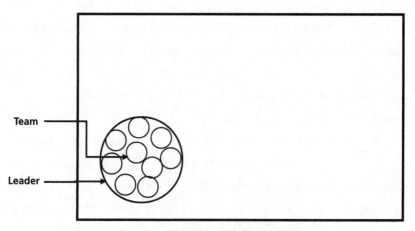

Figure 6. An intelligent individual

don't say what they truly think, but rather what they think the leader wants to hear. They echo his thoughts and anticipate his feelings. There is an absence of rebel ideas.

In effect, they start to migrate toward the alpha, parroting his viewpoint, shrinking their own bandwidth in the process. The cognitive capacity of the team effectively collapses to the parameters of just one brain, as in figure 6. A team of rebels has, through the process of a dominance dynamic, become the social equivalent of a team of clones.

Studies of healthcare have shown that junior members of surgical teams fail to speak up because of fear of the surgeon. The more overbearing the surgeon, the stronger the effect. Remember that leaders are often positioned not merely as powerful, but as smart. How easy is it for a junior member to comfort themselves with the thought that they don't need to speak up because the leader already knows whatever they have to say? How easy when this dovetails with the demands prewired by the evolved psychology of subordination?

From this perspective, the behavior of the engineer on United Airlines 173 begins to make more sense. One can almost sense his thought processes as the fuel runs down, desperately trying to make his concerns known, constrained from doing so by the invisible influence of dominance, frantically seeking to justify his own silence, focusing his frazzled mind on the possibility that the captain already knows the state of the fuel but has come up with a solution.

When the engineer finally made his concerns explicit, it was too late. The team had all the information it needed, but it wasn't communicated. The dividends associated with cognitive diversity—in this case, the utterly prosaic fact of two people focusing on different aspects of a rapidly changing situation—were squandered. The outcome was disaster.

III

Rob Hall reached the summit of Everest at 2:20 p.m. The views were spectacular, the neighboring peaks of the Himalayas seeming like foothills below, the vertex so elevated that it was almost possible to glimpse the curvature of the earth.

Hall was elated. Minutes earlier, when he had been traversing the Hillary Step, he passed Krakauer, one of the members of his team who had already summited, descending. They embraced. Krakauer thanked Hall for masterminding the expedition that had enabled him to fulfil a lifetime ambition. "Yeah, it's turned out to be a pretty good expedition," Hall replied. It would be the last time that Krakauer saw Hall alive.

ROB HALL WAS one of the finest mountaineers in the world. He had summited Everest on four previous occasions. As a leader, he had a deep appreciation of the importance of team cohesion. He had made sure members had gotten to know each other, and that they shared personal stories of what summiting meant to them and their loved ones. It was clear early on that the team, from the climbers to the support staff, were pulling for each other. As Helen Wilton, the base camp manager, put it:

I felt like a part of something great. I really think that to do something with people for a common purpose is a wonderful thing. And to help people achieve their dreams is something that caught me as well. So much emotion and experiences and demands of you happen in such a short space of time: six weeks of intensive living.

This sense of purpose, coupled with the courage of the climbers, would manifest itself in astonishing acts of heroism as the disaster unfolded. But the problem on that fateful day wasn't one of cohesion, as has often been claimed. Nor was it down to the mistakes of any one individual, despite the finger-pointing that followed the disaster. The problem was subtler: dominance dynamics.

Beidleman, a junior guide in the Mountain Madness team whose role would prove key, was in a state of growing anxiety as he stood on the summit of Everest at 2:30 p.m. The turnaround time laid down by Fischer, the team leader, had come and gone, putting pressure on the mobile oxygen supplies. Perhaps Fischer, suffering with illness, wasn't thinking clearly. Perhaps he was unduly influenced by a desire to see his clients reach the summit. What was clear is that Boukreev, the senior guide, had decided to descend alone, imperiling the ratio of guides to clients, creating even deeper anxiety in Beidleman. And yet he didn't challenge the slipping turnaround time, nor the decision by Boukreev to descend.

Why not? On the surface, it seems strange. Intervening was crucial to the safety of the group. It only starts to make sense when you consider status. "Because Beidleman's high altitude experience was relatively limited, his station in the Mountain Madness chain of command was below Fischer and Boukreev," Krakauer writes. "And his pay reflected his junior status." In the months following the disaster, Beidleman made a telling admission, hinting at the steep hierarchy gradient in play on that day. "I was definitely considered the third guide," he said, "so I tried not to be too pushy. As a consequence, I didn't always speak up when maybe I should have, and now I kick myself for it."

But if the dominance dynamics were acute between senior and junior guides, they were even more so between the guides and clients. The clients may not have had the same depth of experience as their leaders, but they had years of altitude climbing under their belts. Moreover, as the climb unfolded, they were gleaning critical information about changing conditions, the physical state of colleagues, and much else besides. Each guide had only one pair of eyes. The team had many.

And yet the very day before the ascent Hall had given a stern speech about the importance not of speaking up, but of mute deference. "[He gave] us a lecture about the importance of obeying his orders on summit day," Krakauer writes. 'I will tolerate no dissension up there,' he admonished, staring pointedly at me. 'My word will be absolute law, beyond appeal. If you don't like a particular decision I make, I'd be happy to discuss it afterward, but not while we're on the hill.'"

Hall gave this speech for what he thought were the best of reasons. He had deeper experience of Everest and was in the best position to make key decisions. But he neglected the fact that his capacity to make wise judgments relied not merely on his own perspective, but those of his team. He rightly stressed the importance of listening to his ultimate judgment, but he didn't realize that it could be fatally compromised without access to the collective intelligence of the group.

The steep hierarchy, which characterized both teams, would manifest itself time and again as the pressure mounted. When Martin Adams, a client on the Mountain Madness team, was inching ever closer to the summit, he noticed something that made his heart beat faster: what looked like wispy clouds below, but which he realized were thunderheads. Adams was a commercial pilot and had long experience of interpreting cloud formation. "When you see thunderheads in an airplane, your first reaction is to get the fuck out of there," he later said.

But he didn't speak up. Neither the guides nor his fellow teammates realized what the clouds meant. They didn't have Adams's experience when it came to subtle variations in vertical cloud patterns. As Krakauer put it, "I was unaccustomed to peering down at cumulonimbus cells from 29,000 feet and I therefore remained ignorant of the storm that was even then bearing down." And yet Adams didn't inform the guides of this critical information.

Dire as the stakes were, the psychology at play was more powerful. The guides had been positioned as leaders. They were the dominant figureheads. The clients had been instructed to obey decisions rather than contribute to them. Speaking up probably didn't cross Adams's mind. The guides, who should have been abandoning the summit attempt and getting everyone down to safety, did no such thing.

A few minutes later, Krakauer, one of the few now on the descent, arrived at the South Summit, keen to get his hands on the supplementary oxygen stashed there. He saw Harris, one of the guides, sorting through a pile of bottles, and expressed his relief that they could now take advantage of the fresh supply. But Harris replied in a curious way. "There's no oxygen here," he said. "These bottles are all empty." But he was wrong. There were at least six full bottles. It is probable that his regulator had become clogged with ice, meaning the bottles registered empty when he tested them.

Either way, Krakauer knew that Harris was mistaken because he had, by now, grabbed a fresh canister and was sucking in the oxygen his body was craving. And yet he scarcely challenged Harris. He knew Harris was mistaken. He knew that oxygen was crucial to his safety, and to the safety of the group, but he didn't push the matter. Instead, he headed down the mountain, as Harris waited at the South Summit to assist the descending climbers.

Why didn't Krakauer speak up? Didn't he care? Was he oblivious to the safety of his teammates? Think back to the testimony of Professor Flin, who said, "Copilots would rather die than contradict the captain." Think back, too, to the engineer on United Airlines 173. Humans are acutely sensitive to hierarchy, even when the stakes are existential. Self-silencing occurs unconsciously.

In perhaps the most revealing passage of his book, Krakauer writes:

My inability to discern the obvious was exacerbated . . . by
the guide-client protocol. Andy and I were very similar in
terms of physical ability and technical expertise; had we been
climbing together in a nonguided situation as equal partners,
it's inconceivable to me that I would have neglected to recognize

his plight. But on this expedition, he had been cast in the role of invincible guide, there to look after me and the other clients; we had been specifically indoctrinated not to question our guides' judgment. The thought never entered my crippled mind that Andy might in fact be in terrible straits—that a guide might urgently need help from me.

This was another interaction that would have fateful consequences. At 4:41 p.m. Hall—who had now summited—radioed base to say that he and his client Doug Hansen were in trouble above the Hillary Step and desperately needed oxygen. Had he known there were fresh bottles waiting at the South Summit, he could have climbed down to retrieve them before reascending. But Harris cut in to say (wrongly) that the bottles were empty. Krakauer's failure to speak up moments earlier meant, in effect, that Hall remained with Hansen above the Hillary Step, desperately trying to drag him down the summit ridge, bereft of bottled oxygen, just minutes before the storm finally hit.

Time and again, information wasn't shared. Critical decisions were made that didn't reflect what the team, as a whole, knew. And yet it is striking when reading the retrospective accounts of the disaster how mystified participants were as to why they hadn't spoken up. Why didn't I share what I knew! Why didn't I voice my concerns! As Krakauer put it in the context of his failure to challenge Harris, "Given what unfolded over the hours that followed, the ease with which I abdicated responsibility—my utter failure to consider that Andy might have been in serious trouble—was a lapse that's likely to haunt me for the rest of my life."

The mistake—looking from the outside in—is to suppose that participants didn't care enough, that they were not sufficiently motivated to help their teammates. It is worth remembering that those who failed to communicate were putting *themselves* in danger. The problem was not motivation, but hierarchy. Key branches in the decision tree were being selected without the combined wisdom of the group. And with each new branching of the decision tree, the climbers—more than thirty people on the mountainside—were being taken, slowly, but inexorably, toward disaster.

By the time the blizzard finally struck, the accumulation of misjudgments would be magnified into the dimensions of a tragedy. "One minute we could look down and see the camp below, and the next minute, you couldn't see it," one later said. As snow started to fall, visibility vanished. Beidleman and Groom, two of the guides, merged into a single team along with seven clients and two Sherpas as they groped their way toward Camp 4. The sound of the wind was deafening. Their eyelids kept getting stuck together, glued by icicles. They had to be ever conscious of pushing too far to the east, which would take them to disaster over the Kangshung Face.

"As you move further and become more disoriented, the entire time you are doing this, the storm, the wind, the snow, the cold, everything, is moving, is crescendoing," Weathers later said. "And now the noise level is starting to overwhelm you, and you have to yell at each other to be heard at all, and we got a sense that we were just being led like sheep." Hopelessly lost in the maw of the storm, they trudged around in circles. Everyone was now out of supplementary oxygen. "It was like pulling out the plug. There was no electricity," one would recall.

Reading about their plight, one is struck by both the growing desperation, but also the astonishing courage. Some collapsed. They were hauled up by teammates. Others talked of giving up. They were talked out of it. When they huddled by a rock, hoping for a break in the clouds, they almost drifted en masse into deadly slumber. "We knew that going to sleep was the wrong thing to do, and it was way too easy to do," Beidleman said.

"You just suck yourself back, you draw yourself back as far as you could into your down suit hood and just close your eyes and take a few breaths, and . . . let go."

When the clouds parted for an instant, giving them a fix on the camp, five were unable to move. Their bodies had stiffened to hyper-rigidity. Those who could still walk stumbled back to the tents, most falling into an exhausted stupor in the folds of their sleeping bags. It was left to Boukreev, who had avoided the storm by descending ahead of the group, to brave the maelstrom alone. Singlehandedly, he dragged three people back to camp, leaving the last two on the rocks of the col. Any further attempt to mount a rescue would probably have killed him. By now, he was frozen stiff.

High up above, Rob Hall was heroically struggling to save Hansen's life, exhausted and almost comatose, as the storm raged around them, dragging him down the knife-edge ridge, both starved of oxygen. When base camp advised Hall to leave his companion, reasoning that this was his only chance to save himself, Hall refused. Harris, aware of the plight of his friends up above, made an astonishing bid for the Hillary Step. He was never seen again.

Fischer, exhausted and probably suffering with illness, died at the southeast ridge. Namba died on the South Col, one of the two who had been left exposed to the elements overnight. She remains in the record books as only the second Japanese woman to reach the seven summits. Weathers survived the night in what is still regarded as the greatest miracle in mountaineering history, staggering into camp the following morning. He was later helicoptered off the mountain, suffering from severe frostbite. His right arm was amputated between the elbow and wrist, and he lost all the fingers of his left hand as well as parts of both feet. His nose was reconstructed with tissue from his ear and forehead. His story is one of the most inspirational in the genre.

Hall continued in his lonely battle to save Doug Hansen in the aerie above the Hillary Step. In his haunting documentary about the expedition, David Breashears—a filmmaker who was at base camp as the disaster unfolded—couldn't help wondering what had happened. "It must have been a desperate struggle as he tried to move Doug along that ridge, only a few feet at a time, so far from the safety of camp . . . And what happened to Doug? Did he still have enough life in him to reach out to Rob and say, 'Don't leave me'? Or did Doug ever look at Rob and say, 'Just go. Save yourself.'"

We will never know the answers to these questions. All that can be said with certainty is that Hall battled to save Hansen until the end, surrendering his own life in a vain attempt to get both men down from the rooftop of the world, just as Harris lost his life after climbing back up the ridge to the Hillary Step, having heard the cry for help from his teammates.

One reads of these actions with a sense of awe. One thrills to their heroism. These individuals were pulling for each other, making sacrifices for each other, risking their lives for each other. Even Boukreev, criticized severely (and, to many, unjustly) by Krakauer for descending to Camp 4 in advance of his team, risked his life not once or twice, but three times, as he dragged stricken comrades back into the tents amid the savagery of the blizzard.

What this calamity shows, and why it is so revealing, is that a team ethic, while precious, isn't sufficient. No amount of commitment can drive effective decision making in a situation of complexity when diverse perspectives are suppressed—when critical information isn't flowing through the social network. By inadvertently creating a dominance dynamic, Hall deprived himself of the very information he needed to make life-and-death decisions as the pressure intensified.

It cost him his life.

IV

Let us leave Everest and examine decision making in the real world. Many of our most important decisions happen in meetings. There are kickoff meetings, town hall meetings, work meetings, board meetings, management meetings, staff meetings, breakfast meetings, off-site meetings, and videoconference meetings. Millions of meetings, daily, around the world.

The logic of holding meetings is sound. Many brains are more effective than one brain—provided these brains are diverse. Over the last two decades, the time spent by managers and employees in collaborative activities has increased by more than 50 percent. But it is here we must confront a sobering, rarely discussed truth. Study after study has revealed the same finding: meetings are catastrophically inefficient. As Leigh Thompson, an academic at the Kellogg School of Management, told me, "Meetings predict terrible outcomes even more powerfully than smoking predicts cancer."

Thompson is a professor of dispute resolution and organizations, and has spent her life studying group judgment. She became interested in human relationships as a teenager when she witnessed her parents' painful divorce. She briefly considered becoming a marriage counselor, but decided that she wanted a broader understanding of human interaction.

As she conducted her research, she quickly noticed dominance dynamics. When one or two people dominate, their authority suppresses the insights of others in the team, particularly the introverts. If the dominant person is the leader, the results are even worse, with people parroting back his or her opinions. Rebel ideas are not expressed. Thompson says, "The evidence suggests that in a typical four-person group, two people do 62 percent of the talking, and in a six-person group, three people do 70 percent of the talking. It gets progressively worse as the group size gets bigger." In fact, this phenomenon is so common that it has spawned a name: *the uneven communication problem.* "Perhaps the most remarkable thing is that the people doing all the talking don't realize they are doing it," Thompson says. "They are adamant that everyone is speaking equally, and that the meetings are egalitarian. The reason is that they often lack self-awareness. So, if you point it out to them, they bristle, and you often get into an escalating conflict."

Thompson's work tallies with the research of Anita Woolley, a psychologist at Carnegie Mellon University, who led an experiment with more than seventy-eight groups assigned to different tasks, from creativity to decision making. The researchers assumed that the teams with highest aggregate IQ would perform the best.

Two factors turned out to be more important than IQ. Teams in which members spoke for similar amounts of time performed far better than those dominated by one or two voices—researchers call this *conversational turn taking.* The second factor was social perceptiveness: teams performed better when populated by people who could read each other's moods and meanings. These teams tended to have more women, who, on average, have higher levels of social intelligence.

These results are compelling. When one person dominates, the insights of other team members are crowded out. They do not get a chance to contribute. Social perceptiveness facilitates the flow of information by ensuring that perspectives are not just expressed, but understood. It sometimes takes emotional intelligence not only to hear what someone is saying but also to grasp what they mean. As Woolley says, "Having really smart people in the group wasn't itself enough to make a smart group. . . . Rather, the thing that drove collective intelligence was the way the group interacted. When they interacted effectively, they exceeded the capability of individual members."

The problem is that at most meetings communication is dysfunctional. Many people are silent. Status rigs the discourse. People don't say what they think but what they think the leader wants to hear. And they fail to share crucial information because they don't realize that others lack it. In one experiment, a team was tasked with hiring a manager from three candidates. The researchers rigged the attributes of the candidates to ensure that one was far better than the other two. Better qualified, superior characteristics, a better fit. The researchers then provided candidate information to the members of the hiring team—but with a twist. Each team member received an information pack with only a subset of information about the three candidates. The team, as a whole, had all the information, but each member only had a portion of it. This meant that the correct decision could only emerge if each person shared what they knew. What happened? The teams failed miserably. Almost none made the right choice.

This is significant because this example describes what most teams experience. Each person has something useful to contribute (otherwise, why would they be on the team?), but instead of this being harnessed as part of a group decision, one member of the group, acting on limited information, expresses a preference, skewing the entire dynamic. People start to share the information that corroborates that view, and subconsciously withhold information that might call it into question. Diversity of thought vanishes. This is called an *information cascade*.

When team members in a control group were each given access to all the information, rather than each receiving a subset of the information, they made the right choice every time. As psychologist Charlan Nemeth puts it, "Group processes by and large conspire to suppress the very diversity of viewpoints that we seek."

This takes us back to one of the insights of the previous chapter. You'll remember that in prediction tasks, taking the average of independent estimates can lead to remarkably accurate judgments. This is the wisdom of the crowd. This outcome has been found in multiple settings from economic forecasting to the experiment in which students were asked to estimate the length of the London Underground.

But now suppose that the students guessing the length of the Underground had guessed not independently (by writing on slips of paper) but sequentially. The first person announces her estimate, followed by the person sitting next to her, and so on. The first guess is now not merely an estimate but also a signal to everyone else. The next person may copy that guess or lean toward it, thus influencing the third person. The errors are no longer cancelling, they are correlating.

This is another example of an information cascade, and much of its force is explained by interpretation. When two or more people lean toward the same answer, it is easy to assume they arrived at it independently. This amplifies its persuasive power, causing others to lean toward it, too. This is where fads, stock market bubbles, and other bandwagon effects come from. Crowds are not always wise. They can become dangerously clone-like.

These cascades can happen at a purely social level, too. Studies by the psychologist Solomon Asch have shown that people often lean toward the answers of others, not because they believe them to be correct, but because they don't want to appear rude or disruptive by disagreeing. And this brings us back to dominance, which can be thought of as a social dynamic that vastly magnifies the dangers of information and social cascades. After all, if we find it difficult to contradict the opinion of strangers, how much more difficult is it to contradict the opinion of the leader?

With prediction tasks, cascades can be avoided by taking independent estimates, but this is not possible with most other decisions. With problem-solving, policymaking, and the like, where debate and discussion are needed to hear and test different perspectives, we cannot avoid meetings—which is why we need to understand their defects.

By compounding each other's errors, rather than correcting them, teams can become increasingly confident about objectively terrible judgments. As

Cass Sunstein and Reid Hastie, two experts on group decision making, put it, "Much of the time, groups blunder not in spite of group deliberation, but because of it. After deliberation, companies, labor unions, and religious organizations often make disastrous decisions. The same point holds for governments."

It is a curious irony. We spend much of our lives building up individual expertise. We spend years at school, then at university, then we undergo apprenticeships or on-the-job training, gradually attaining knowledge, insight, and understanding. We then make the biggest decisions in forums that make us collectively dumb.

V

Early in its history, Google decided to get rid of all managers. The founders wanted an entirely flat structure. They had noted the mounting evidence on the defects of hierarchies and wanted to counteract them. It didn't work. As the psychologists Adam Galinsky and Maurice Schweitzer put it in their book *Friend and Foe*:

> Early on, founders Larry Page and Sergey Brin conducted
> what they thought would be a revolutionary experiment: they
> eliminated managers and created a completely flat organization.
> The experiment was indeed eye-opening but only because it was
> a failure. The lack of hierarchy created chaos and confusion, and
> Page and Brin quickly realized that Google needed managers to
> set direction and facilitate collaboration. As they learned, even
> Google needs some hierarchy.

Other studies have found similar results. One led by Eric Anicich of Columbia University examined fashion houses between 2000 and 2010, evaluating performance with the industry standard: the French trade magazine *Journal du Textile.* The takeaway was clear: fashion houses with codirectors were rated as less creative than those with single directors. As Galinsky and Schweitzer put it, "Coleadership can kill ideas because it creates uncertainty over who is in charge."

Groups typically need a leader, otherwise there is a risk of conflict and indecision. And yet the leader will make wise choices only if they gain access to the diverse views of the group. But how can an organization have hierarchy and information sharing, decisiveness *and* diversity? This is the question that has dominated management books for decades, typically with hierarchy and diversity portrayed as inherently conflicting. The idea has been to shift the hierarchy gradient such that you get a bit of dominance along with a bit of diversity.

But this analysis overlooks a key point. Hierarchy is an inevitable aspect of most human groups. We cannot ignore it. But our species, uniquely, doesn't have just one form of hierarchy. We have two.

FROM 1906 TO 1908, A. R. Radcliffe-Brown, the great British anthropologist, lived among the hunter-gatherers of the Andaman Islands. While there, he noted an anomaly. Some individuals attained influence in the community, and seemed to command deference, and yet did not engage in dominance behaviors. Their status seemed to be constructed on something else. He writes:

> There is another important factor in the regulation of social life, namely the respect for certain personal qualities. These qualities are skill in hunting . . . generosity and kindness, and freedom from bad temper. A man possessing them inevitably acquires a position of influence in the community. Others . . . are anxious to

please him by helping him in such work as cutting a canoe or to join him in hunting parties or turtle expeditions.

People found themselves in positions of leadership, then, not by threatening or intimidating subordinates, but by gaining their respect. Hierarchies organically developed not through dominance, but what seemed like a distinct mechanism. A mechanism that, to Radcliffe-Brown, was stable and consistent, and had its own suite of postures, behaviors, and expressions. When Radcliffe-Brown's account was published, his descriptions could have been taken as an idiosyncrasy of a particular tribe, but other anthropologists realized that they had observed similar dynamics in other groups, but without noting their significance. It had been observed in the Aborigines, in the Tsimané in the Bolivian Amazon, the Semai people of Malaysia, and others.

Anthropologists living in the West had noticed them, too. They had seen leaders—formal or informal—who did not demand respect from subordinates, but who earned it, whose status was signaled not by aggression, but wisdom, whose actions tended not to intimidate, but to liberate.

Sure enough, when psychologists started to look for these dynamics among strangers in a lab, they detected the emergence of a different kind of social hierarchy alongside dominance. It was not just observable in the way people behaved when solving problems, but could be glimpsed by outsiders. And this form of social status was present in different cultures, tribes, and nations. To distinguish this form of social status from dominance, anthropologists gave it a different name: *prestige*. Joseph Henrich, the Harvard academic who cowrote the most widely cited paper on prestige, has said, "Both dominance and prestige are clearly discernible, with predictable patterns of behavior, postures, and emotions. And they provide different routes to status."

We can see the different features in the following table, taken from the work of Henrich and Maner:

STATUS FEATURES	DOMINANCE	PRESTIGE
History	Ancient, dating back to at least common ancestors of humans and other nonhuman primate species	Unique to humans; emerged when humans lived in relatively small hunter-gatherer communities
Source of deference	Deference is demanded and is a property of the actor	Deference is freely conferred and is a property of the beholder
Mechanisms of influence	Coercion, intimidation, aggression, manipulation of reward and punishment	True persuasion, respect, liking, social modeling
Role of social bonds	Opportunistic and temporary use of social coalitions as a means of attaining social rank	Creation of authentic and lasting relationships with other group members
Personality	Narcissistic, high in hubristic pride	Authentic pride
Attention by lower status	Tracking of higher-ups, avoidance of eye contact, and no staring	Directing of attention to and gazing at higher-ups, watching and listening
Proximity management	Avoidance of higher-ups; keeping distance to avoid aggression	Approach to higher-ups, maintenance of proximity
Display by lower status	Diminutive body position, shoulder slump, crouching and gaze aversion	Attention to prestigious, open-body position
Display by higher status	Expansive body position, expanded chest, wide stance, arms wide	Similar to dominance but more muted, less expansive use of space
Social behavior	Aggression, self-aggrandizement, egocentrism	Prosocial, generous, cooperative

Why did prestige evolve in humans? Why would a prestigious person share wisdom in the first place? Wouldn't it be more advantageous to keep it to themselves? There are many facets to the explanation, but the key point is simple. Remember that dominant individuals are mimicked out of fear. Prestigious individuals, on the other hand, are followed out of freely bestowed respect. They are role models.

This means, in turn, that their generosity toward others is likely to be copied, tilting the entire group in a more cooperative direction. The prestigious individual may have conferred an advantage on someone else, but she benefits from the broader adoption of generosity across the group. This is particularly important if helping each other amplifies the overall payoff—so-called positive-sum environments. This is precisely the historical context in which prestige first developed.

Dominance hierarchies have different internal dynamics. Given that moving up the hierarchy depends on someone else moving down, it tends to accentuate zero-sum behavior. In other words, politicking, backstabbing, and quid pro quos, along with constant vigilance about internal competition. Chimpanzees, for example, are masters of strategic coalitions to thwart internal rivals, since "rank competitions are typically won by those who have the confidence to enter into potentially violent interactions, and by those with the social support to back up their advances."

This explains why prestigious human leaders tend not to bare their teeth or wave their arms. On the contrary, they use self-deprecation as a rhetorical device to signal a different dynamic. They explain their ideas thoroughly, because they know that colleagues who understand and endorse them are more likely to execute them with judgment and flexibility. They listen to those around them, because they recognize that they are not too smart to learn from others.

Maner argues that dominance and prestige should be considered not as distinct personality types, but more as techniques. Dominance, as a technique, retains its logic today. When a decision has been made, and there is no turning back, dominance makes sense. Leaders need to galvanize their teams to get the job done. Dissent and diverse opinions are a distraction. But when evaluating as opposed to executing decisions, or coming up with new ideas, dominance tends to collapse under the weight of its own contradictions. This is where a prestige dynamic is crucial. People need to speak up, to offer rebel ideas, safe from the retribution of a leader who interprets such contributions as a threat.

This analysis dovetails with one of the most influential concepts in modern organizational research: *psychological safety.* When an environment is psychologically safe, people feel they can offer suggestions and take sensible risks without provoking retaliation. The connection between prestige-oriented leadership and psychological safety should be obvious; let us focus on the role of empathy.

Dominant leaders are, by definition, punitive. This is how they win and sustain power. They are also less empathetic. They don't feel that they need other people, so don't tend to take their perspectives or read their emotions. Prestige-oriented leaders, on the other hand, recognize that wise decisions depend on the input of the group, and so are highly attuned to what others are thinking and saying. This behavior strengthens trust. "Prestige is associated with higher empathy and information sharing," Maner says. "This boosts collective intelligence."

A major investigation by Google, which sought to identify why some teams perform better than others, found that psychological safety was the single most important factor driving success, a result that has been widely replicated. "Psychological safety was far and away the most important of the dynamics we found," their report stated. "And it affects pretty much every important dimension we look at for employees. Individuals on teams with higher psychological safety are less likely to leave Google, they're more likely to harness the power of diverse ideas from their teammates, they bring in more revenue, and they're rated as effective twice as often by executives."

The irony is that most professional environments lack psychological safety. In one study examining retail and manufacturing, employees who frequently offered new ideas and concerns were significantly less likely to receive pay raises or promotions. The penalties were even higher for women, where speaking up can violate gender stereotypes. This can be exacerbated still further for women who are also members of an ethnic minority, something described by psychologists as *double jeopardy*. Nemeth writes, "We are afraid of the ridicule or rejection that are likely to come from dissenting. We hesitate. We put our heads down. We are silent. Not speaking up, however, has consequences."

And this is why a new generation of leaders has shifted toward a prestige approach. This is a key aspect of how General Stanley McChrystal turned around the battle against Al Qaeda after the invasion of Iraq and how Satya Nadella helped rebuild the fortunes of Microsoft—by recognizing the limitations of dominant leadership. Soon after coming to power, Jacinda Ardern, the prime minister of New Zealand, said, "It takes power and strength to be empathetic." Respect was not demanded by these leaders, but volunteered by those they led.

"Leaders often worry that inviting other views—particularly those that disagree with them—might undermine their authority," Nadella told me. "They are wrong. Most people feel more committed when they are given the opportunity to make a contribution. It strengthens motivation, boosts creativity, and increases the potential of the entire organization." Maner says, "There is a time and place for prestige, and a time and place for dominance. Wise leaders are able to pivot back and forth between the two. When executing a plan, dominance can be crucial. But when deciding on a new strategy, or predicting the future, or finding new innovations, you need to hear diverse perspectives. This is where dominance can be disastrous."

In addition to creating a culture of psychological safety, cutting-edge organizations have also started to introduce specific mechanisms to safeguard effective communication. One of the most celebrated is the "golden silence" of Amazon. For more than a decade, meetings at the tech giant have started not with a PowerPoint presentation or banter, but total silence. For thirty minutes, the team reads a six-page memo that summarizes, in narrative form, the main agenda item.

This has a number of effects. First, it means the proposer has to think deeply about their proposal. As Jeff Bezos, founder and longtime CEO of Amazon, put it, "The reason writing a 'good' . . . memo is harder than 'writing' a twenty-page PowerPoint is because the narrative structure of a good memo forces better thought and better understanding of what's more important than what." And later: "It has real sentences, and topic sentences, and verbs, and nouns—it's not just bullet points."

But there is a deeper reason why this technique is powerful: it commits people to deciding what they think *before* learning the opinions of others. They have the space to bring their diverse ways of thinking, reasoning through the weaknesses and strengths of the proposal, before discussion. This reduces the risk that diverse perspectives will fail to surface. And even when discussion *does* start, the most senior person speaks last, another technique that protects diversity of thought.

In a post on LinkedIn, Brad Porter, a vice president at Amazon, described these simple mechanisms as among the most important strategic advantages of one of the world's most successful companies. "I don't think I'm revealing Amazon's secret sauce by describing the process," he said. "Where I run some risk of revealing too much is by telling you that

Amazon absolutely runs better, makes better decisions, and scales better because of this particular innovation."

Another technique is brainwriting. Like brainstorming, this is a way of generating creative ideas, but instead of stating the ideas out loud, team members are asked to write them down on cards, which are then posted on a wall for the rest of the group to vote on. "This means that everyone gets a chance to contribute," Thompson told me. "It means that you gain access to the output of every brain, rather than just one or two."

Thompson suggests that brainwriting should have just one rule: nobody is allowed to identify themselves with their written contributions. The marketing director should not offer a "tell" by referring on the card to a client associated with him. "This is crucial," Thompson says. "By anonymizing the contributions, you separate the idea from the status of the person who came up with it. This creates a meritocracy of ideas. People vote on the quality of the proposal, rather than the seniority of the person who suggested it, or to curry favor. It changes the dynamic."

After voting on the ideas, groups are typically divided into four-person teams to "take them to the next level," combining ideas, or sparking new insights. "Using this iterative technique, brainwriting can be woven into interactive team meetings in a way that engages everyone," Thompson says. When brainwriting is put head-to-head with brainstorming, it generates twice the volume of ideas, and also produces higher-quality ideas when rated by independent observers. The reason is simple. Brainwriting liberates diversity from the constraints of dominance dynamics.

Ray Dalio has built one of the most successful hedge funds with similar methods. Bridgewater Associates operates according to more than two hundred behavioral "principles," but the key theme can be summarized in one phrase: the expression of rebel ideas. He calls it *radical transparency*. The culture is one where people are not fearful of expressing what they think; it is a duty. As Dalio put it in an interview with Adam Grant, "The greatest tragedy of mankind comes from the inability of people to have thoughtful disagreement to find out what's true."

At another company, every person invited to a meeting is asked to submit a one-pager on their views. This is what you might call the price of attendance. These one-pagers are then shuffled, handed around the table, and read out in random order. This is another way of separating a perspective from the status of the person who proposed it.

These techniques may seem like gimmicks, but they share an effective underlying pattern. They protect cognitive diversity from the dangers of dominance.

VI

In 2014, Anicich, now at the University of Southern California, collected data from over 30,625 Himalayan climbers from 56 nations on over 5,100 expeditions. It was the largest analysis of high-altitude mountaineering ever conducted. The researchers were interested in one issue above all else: Do dominance hierarchies lead to a higher probability of disaster?

They couldn't measure the hierarchies of the teams directly, given that the climbers were dispersed around the world, and many of the expeditions had occurred years previously. But they did the next best thing. They examined the nations they hailed from. Some cultures are deferential to authority figures, and are, on average, less likely to speak up. Other cultures tolerate and even encourage speaking up to those in positions of leadership.

Would these small national differences show up in the data? Would they emerge in the number of fatalities? As Anicich probed the evidence, the answer was clear: teams with more dominant hierarchies are "significantly more likely to die." This finding did not apply to solo expeditions. It was only teams from more hierarchical nations that had the problem, which shows that it wasn't about the skill of individual climbers but how they interacted. Galinsky, one of the coauthors, writes:

In cultures that are hierarchical, decision making tends to be a top-down process. People from these countries are more likely to die on difficult mountain climbs because they are less likely to speak up and less likely to alert leaders to changing conditions and impending problems. By not speaking up, these climbers preserved order but endangered their own lives. Importantly, we isolated the role of *group* processes by showing that the higher fatality rate occurred for group, but not solo, expeditions. It was only when a group of individuals had to communicate effectively that hierarchical cultures produced disaster.

This finding, published in the *Proceedings of the National Academy of Sciences,* is significant enough on its own. But it becomes compelling as an explanation for the Everest disaster when corroborated by the evidence from Google, anthropological data, controlled lab studies, and more. As Galinsky put it, "The Himalayan context highlights a key feature that creates complex decisions: a dynamic and changing environment. When the environment can change dramatically and suddenly, people have to adapt and come up with a new plan. In these cases, we need everyone's perspective brought to bear and hierarchy can hurt by suppressing these insights."

It is worth reiterating that none of this invalidates the notion of hierarchy. Most teams function better with a chain of command. Hierarchy creates a division of labor, where leaders can focus on the big picture while others grapple with the detail. It also ensures that teams can coordinate their actions. If there is no hierarchy, team members might constantly argue about what to do next. This can be disruptive and dangerous.

But the real choice is not between hierarchy and diversity, but about how to gain the benefits of both. As Galinsky put it:

> In complex tasks, from flying a plane to performing surgery
> to deciding whether a country should go to war, people need to
> process and integrate a vast amount of information while also
> imagining myriad possible future scenarios. . . . To make the best
> complex decisions, we need to tap the idea from all rungs of the
> hierarchical ladder and learn from everyone who has relevant
> knowledge to share.

LET US CONCLUDE our analysis of hierarchy with perhaps the deepest irony of all. One emphatic finding from psychological research is that humans dislike uncertainty and the sense that we lack control over our lives. When faced with uncertainty, we often attempt to regain control by putting our faith in a dominant figurehead who can restore order. This is sometimes called *compensatory control.*

Look at the rise of authoritarian states in times of economic uncertainty, such as in Germany and Italy after the chaos of the First World War. A study led by Michele Gelfand of the University of Maryland analyzed more than thirty nations and found that they responded to external forces that threated certainty or security by reaching for steeper political hierarchies.

This has religious implications, too. One study analyzed church membership in the United States over two periods, one characterized by job security (1920s) and one characterized by intense uncertainty (1930s). The researchers then divided churches into two categories: hierarchical churches with many layers of authority (Roman Catholic Church, Church of Jesus Christ of Latter-Day Saints) and nonhierarchical with few layers (Protestant Episcopal Church, Presbyterian, and so on).

Sure enough, when the economy was rosy, people were far more likely to join nonhierarchical churches. When jobs were insecure, and people felt that they lacked control over their lives, on the other hand, they tended to favor the hierarchical churches. They compensated for their feelings of insecurity by putting their faith in theologies with higher levels of theistic power and control.

If this seems a little abstract, think of the last time you experienced heavy turbulence on a plane. Did you whisper a silent prayer? This is another classic manifestation of compensatory control. In the teeth of uncertain events, we reintroduce certainty by imputing power to God, or fate, or some other omnipotent force. It may not make the plane any safer (depending on your religious outlook), but it makes us *feel* a little more secure.

This happens in organizations, too. When a company faces external threats or economic uncertainty, its shareholders are significantly more likely to appoint a dominant leader. Within organizations, too, dominant individuals tend to rise more rapidly during times of uncertainty. The strong voice, the authoritarian personality, provides reassurance for the loss of control we collectively feel.

This leaves us with a dangerous paradox. When the environment is complex and uncertain, this is precisely when one brain—even a dominant brain—is insufficient to solve the problem. It is precisely when we need diverse voices to maximize collective intelligence. Yet this is precisely when we unconsciously acquiesce in the dubious comfort of a dominant leader. Dominance is not just about the character of leaders, then, but often linked to the silent wishes of those who constitute a team, organization, or nation. Leaders who might naturally favor a prestigious style of leadership can find themselves morphing toward dominance when the team starts to lose control of events—with disastrous consequences.

ROB HALL WAS AN ADMIRABLE MAN. The more one reads about him, the more one understands why he was regarded with such awe. One obituary, written just after the Everest disaster, captured his heroism: "Crippled by frostbite, running out of oxygen, and stranded without food, fluid, or shelter, he . . . died that night. . . . The fact that he died while trying to save an exhausted client confirmed his status as the world's most respected leader of commercial Himalayan expeditions."

Hall was not naturally dominant. He was open and inclusive, a man loved by almost all who knew him. The problem is that he had come to believe that a dominant style of leadership would prove an asset on one of the most challenging climbs of his career—and was encouraged in that view by a team experiencing deep anxiety about the sheer volatility in the Death Zone. These unconscious dynamics play out in organizations, charities, unions, schools, and governments in millions of ways, every day, all over the world, but in the high-stakes circumstances of a Himalayan climb, they would have fatal consequences.

In his last radio communication to base camp, amid the storm on the southeast ridge, Hall was told by his comrades that they would patch him through to his wife, Jan, in New Zealand, seven months pregnant with their first child. Hall asked for a moment to steady himself. He knew that he was now beyond hope, but didn't want his deteriorating condition to cause grief to his beloved. "Give me a minute," he said. "My mouth's dry. I want to eat a bit of snow before I talk to her."

Finally, with his mouth moistened, Hall spoke: "Hi, my sweetheart," he said. "I hope you're tucked up in a nice warm bed. . . ."

"I can't tell you how much I'm thinking about you," Jan replied.

"I'm looking forward to making you completely better when you come home. . . . Don't feel you're alone. I'm sending all my positive energy your way."

More than four hundred feet above the South Summit, his friends Doug Hansen and Andy Harris dead, and with the blizzard still raging around him, Hall uttered his last words:

"I love you. Sleep well, my sweetheart. Please don't worry too much."

4

INNOVATION

I

David Dudley Bloom was, by any reckoning, a remarkable man. Born in Pennsylvania on September 20, 1922, he served in the U.S. Navy during the Second World War and, according to some accounts, became the youngest commanding officer in the fleet, taking charge of USS *Liberty* in December 1944 during the New Guinea campaign. At the time he was just twenty-two.

After leaving the army in 1945, he worked in different jobs—including as a clerk in a law firm and a buyer in a department store—before becoming director of product research at American Metals Specialties Corporation

(AMSCO), a small toy manufacturer. Perhaps because of his experiences of war, he sought to move the company away from military-themed toys—such as guns, rifles, and soldiers. As he put it in an interview in the 1950s, "If we teach our children war and crime, we haven't much of a future to look forward to."

His first big idea was a "magic milk bottle," the milk seeming to disappear from the bottle as it was turned upside down. He also came up with miniature consumer products, such as kitchen utensils, so that children could pretend to be chefs.

But it wasn't until 1958 that Bloom hit on the idea that should have changed his life. He had left AMSCO a few months earlier to work for the Atlantic Luggage Company of Ellwood City, Pennsylvania, where he had been offered the job of director of product development for a popular line of travel luggage. He was struck by a thought: Why do suitcases, heavy and burdensome, and which had been partly responsible for his own back pain, not have wheels? Wouldn't wheels make them easier to move around? Wouldn't wheels remove the need for expensive porters? Wouldn't wheels alleviate the sense of foreboding when arriving in transit, the ever-more-frequent shifting from one hand to the other as one trudged from place to place, until one merely had a choice between continuing pain in one arm, or new pain in the other? More generally, wasn't this a perfect solution for a world moving in the direction of mass travel (the fact that so many more people were traveling than ever before)?

He took a prototype of his idea—a suitcase attached to a platform with casters and a handle—to the chairman of the Atlantic Luggage Company. Bloom was expectant, almost exultant. The product would be cheap to make, would tie in with the company's existing designs and distribution channels, and seemed like the most surefire thing in the history of the sector, enabling them to dominate a multibillion-dollar global market.

The chairman's reaction? He described it as "impractical" and "un-wieldy." "Who'd want to buy luggage on wheels?" he scoffed.

IN 2010, IAN MORRIS, a British archaeologist and historian, completed a seminal study into the history of innovation. He was nothing if not thorough. He examined development from 14,000 BC to today, carefully tabulating the consequences of every leap forward.

The major episodes were not difficult to spot. The domestication of animals. The birth of organized religion. The invention of writing. Morris noted that each of these events had eloquent advocates when it came to the following question: Which single change had the greatest impact on humanity? Morris wanted an objective answer, so he painstakingly quantified each breakthrough in social development. This he defined as "a group's ability to master its physical and intellectual environment to get things done," an idea that correlates closely with economic growth.

His data shows that all the previously mentioned innovations did indeed influence social development. The line gradually slopes upward over the course of the millennia. But one innovation had an impact beyond any other, shifting the curve from near horizontal to near vertical: the Industrial Revolution. Morris writes, "The Western-led takeoff since 1800 made a mockery of all the drama of the world's earlier history." Erik Brynjolfsson and Andrew McAfee, two professors at the Massachusetts Institute of Technology's (MIT) Sloan School of Management, concur: "The Industrial Revolution ushered in humanity's first machine age—the first time our progress was driven primarily by technological innovation—and it was the most profound time of transformation our world has ever seen."

But there was one anomaly in this picture. When historians zoomed in on this transformation and could see the granularity of the curve, they noticed something odd. The second phase of the Industrial Revolution came with electrification in the late nineteenth century. This meant that electrical motors could replace the older, less efficient steam engines. It created a second surge in growth and productivity, the consequences of which we are still living with today.

Except for one thing. This surge was curiously delayed. It didn't happen instantly, but seemed to pause, pregnant and static, for around twenty-five years before taking off. Perhaps the most curious thing of all is that many of the most successful corporations in the United States, which were in the perfect position to benefit from electrification, did nothing of the kind. On the contrary, many went bust. They were on the verge of victory, and snatched defeat from its jaws.

Electricity, it is worth noting, offered huge dividends, not just in terms of power but in the redesign of the manufacturing process itself. In a traditional factory, machines were positioned around water and, later, the steam engine. They clustered in this way out of necessity. The production process was umbilically linked to the sole source of power, with the various machines connected via an elaborate—but often unreliable—set of pulleys, gears, and crankshafts.

Electrification meant that manufacturing could break free of these constraints. Electric motors do not suffer major reductions in efficiency when reduced in size, so machines could have their own source of power, allowing the layout of factories to be based around the most efficient workflow of materials. Instead of a single unit of power (the steam engine), electricity permitted group power. It is as obvious an advantage as adding wheels to a suitcase. As McAfee and Brynjolfsson put it, "Today, of course, it's completely ridiculous to imagine doing anything *other* than this; indeed, many machines now go even further and have multiple electric motors built into their design. . . . It's clear that intelligent electrification made a factory much more productive than it could otherwise be."

Electrification, then, was a gift from the gods, an opportunity for the companies that dominated U.S. production to increase their efficiency and profits. They had the existing plants. They had the existing machines. And they now had a technology—electricity—to increase their efficiency, streamlining their operations and opening up new streams of growth.

And yet they didn't do anything of the kind. In a move eerily reminiscent of the early luggage companies that turned their backs on wheels, many stuck to unit drive. Instead of streamlining their factories, they dumped a large electric motor in the middle of the factory, as if it were a substitute steam engine. In doing so, they completely—almost inexplicably—missed the point. This would prove catastrophic. The economist Shaw Livermore found that more than 40 percent of the industrial trusts formed between 1888 and 1905 had failed by the early 1930s. A study by Richard Caves, another expert in economic history, found that even those that managed to remain in existence shrank by over a third. It was one of the most brutal periods in industrial history. This is a pattern that repeats endlessly. Organizations in the perfect position to win manage, against all odds, to lose.

When Bernard Sadow, another enterprising executive, brought the idea of wheeled suitcases to market, the secondary beneficiaries—the department stores—wanted to throw the new profits away, too. Sadow had come on the idea in 1972 while struggling with two heavy suitcases through an airport when returning from a family holiday in Aruba. "It just made sense," he would later say.

And yet when he took the idea to the New York stores, who had so much to gain in new sales, he was knocked back. It was like the experience of Dudley Bloom all over again. "Everybody I took it to, threw me out—from Stern's, Macy's, A&S, all the major department stores," Sadow has said. "They thought I was crazy, pulling a piece of luggage. . . . At this time, there was this macho feeling. Men used to carry luggage for their wives."

It was only when he got to see Jerry Levy, a vice president at Macy's, that he managed to secure a deal. Levy called in Jack Schwartz, the original Macy's buyer who had rejected the bag a few weeks earlier, and urged him to buy it. Customers, for their part, experienced no resistance to the new invention. "The people accepted it immediately," Sadow said. "They could see what it was doing. It took off. It was terrific."

As for the history of electrification, it defies logic even more than the history of wheeled suitcases. The executives of the industrial trusts were far from unintelligent. Many were among the early wave of professional managers, handpicked for their keen minds. And yet they turned a gilt-edged opportunity for growth into a disaster on an epic scale. As McAfee and Brynjolfsson put it, "In the first decades of the twentieth century, electrification caused something close to a mass extinction in U.S. manufacturing industries."

II

We have seen how diversity can enhance collective intelligence in everything from climbing mountains to making policy to cracking secret codes. In this chapter, we are going to look at the context with arguably the most dramatic implications: innovation and creativity. We will examine the big picture. Why are some institutions and societies more innovative than others? How can we harness diversity to boost economic prosperity? But first, we will focus on individuals. Why do some people embrace change while others fear it? Why do some master the art of reinvention, while other seem stuck with the status quo?

Experts on innovation often distinguish between two different kinds. On the one hand, there are the directed, predictable steps that take one deeper into a given problem or specialism. Think of James Dyson patiently tweaking the design of his vacuum cleaner, learning more about the separation of dust from air as he adjusted the dimensions of his famous cyclone. With each new prototype, he learned ever more about separation efficiency. With each new step, he gained deeper knowledge of this small segment of science. With each new experiment, he got ever closer to a functional design. This kind of innovation is sometimes called *incremental*. It neatly denotes the idea of knowledge deepening within well-defined boundaries.

The other kind of innovation is embodied in the two examples just discussed. This is sometimes called *recombinant innovation*. You take two ideas, from different fields, previously unrelated, and fuse them together. A wheel with a suitcase. A new form of power generation with a reformed manufacturing process. Recombination is often dramatic, because it bridges domains, or breaks down silos altogether, opening up new seams of possibility.

The logic of these two forms of innovation has an echo in biological evolution. We might think of incremental innovation as somewhat like natural selection, small changes occurring in each generation. Recombinant innovation is rather more like sexual reproduction, genes from two distinct organisms joining together. And while both are important, the science writer Matt Ridley has persuasively argued that we have long underestimated the power of recombination. He writes:

Sex is what makes biological evolution cumulative, because it brings together the genes of different individuals. A mutation that occurs in one creature can therefore join forces with a mutation that occurs in another. . . . If microbes had not begun swapping genes a few billion years ago, and animals had not continued doing so through sex, all the genes that make eyes could never have got together in one animal; nor the genes to make legs or nerves or brains. . . . Evolution can happen without sex, but it is far, far slower. And so it is with culture. If culture consisted simply of learning habits from others, it would soon stagnate. For culture to turn cumulative, ideas need to meet and mate. The "cross-fertilization" of ideas is a cliché, but one with unintentional fecundity. "To create is to recombine," said the molecular biologist François Jacob.

Ridley has a neat phrase for recombinant innovation: "ideas having sex."

History has thrown up plenty of examples of recombinant innovation, such as the printing press, which fused an existing method for pressing wine with various other features such as soft metals to create block techniques and movable type. Recombinant innovations have always coexisted alongside the incremental innovation expressed in the continual modification of existing ideas. But something has happened over recent years that has escaped the notice of many people, indeed many scientists. The balance between incremental and recombinant innovation has started to tilt dramatically. Recombination has become the dominant force of change, not just in science, but in industry, technology, and beyond.

To get a sense of this dominance, consider a study led by Brian Uzzi, professor at the Kellogg School of Management. He looked at 17.9 million publications across 8,700 journals in the Web of Science, the world's largest repository of scientific knowledge. That is pretty much every paper written in the last seventy years. He was looking for patterns. What creates great science? Where are the hot ideas?

What did he find? The papers with the most impact were those that had what the researchers called *atypical subject combinations*, that's to say, papers that bridged across traditional boundaries. These papers were blending, say, physics and computation, or economics and network theory, or psychology and evolutionary biology. This is the scientific equivalent of "ideas having sex." These papers broke through the conceptual walls between subjects and thought silos, creating new ideas and possibilities.

As Uzzi put it, "Many of these novel combinations are really two conventional ideas in their own domains. You're taking well-accepted ideas, which is a wonderful foundation—you need that in science. But when you put them together: wow! That's suddenly something really different." Classic examples of recombinant science include Marie Curie, who (among other things) brought together radioactivity and neoplasms, and Kary Mullis, who fused existing techniques to generate a new way of replicating DNA sequences. Curie would go on to win the Nobel Prize twice.

This isn't just about science, however. The U.S. Patent and Trademark Office has broad categories such as utility patents (the light bulb), design patents (the Coke bottle), and plant patents (hybrid corn), with 474 technology classes and 160,000 codes. In the nineteenth century, most patents were classed by a single code. The majority of innovations were confined to a specific silo. They were typically the product of incremental innovation. Today, the number of patents classed by a single code has dropped to just 12 percent. The vast majority of patents now reach *across* traditional boundaries and codes. As Scott Page, professor of complex systems at the University of Michigan, Ann Arbor, says, "The data reveal the value of combining diverse ideas and an unmissable trend toward recombination as a driver of innovation."

The link between recombinant innovation and diversity should be obvious. Recombination is about cross-pollination, reaching across the problem space, bringing together ideas that have never been connected before. We might call these *rebel combinations*: merging the old with the new, the alien and the familiar, the outside and the inside, the yin and the yang.

This trend is not slowing down but accelerating in the computer age, with its vast networks. Think of Waze. This app is classically recombinant, combining a location sensor, data transmission device, GPS system, and social network. Or take Waymo, the self-driving car technology company, which brings together the internal combustion engine, fast computation, a new generation of sensors, extensive map and street information, and many other technologies.

Indeed, almost all tech innovations connect disparate ideas, minds, concepts, technologies, data sets, and more. This pattern applies to Facebook (which connected an existing web infrastructure with tools to build digital networks and share media) and Instagram (which linked Facebook's basic concepts with a smartphone application replete with photo filters) and beyond. Recombination is the leitmotif of digital innovation. With each new combination, fresh combinations loom into the terrain of what the biologist Stuart Kauffman calls the *adjacent possible*. New prospects open up, new vistas come into view. "Digital innovation is recombinant innovation in its purest form," Brynjolfsson and McAfee write. "Each development becomes a building block for future innovations. . . . Building blocks don't ever get used up. In fact, they increase the opportunities for future recombinations."

But this leaves us with a critical question. Why do some people grasp the opportunities of recombination, while others seem blind to its potential? In the examples of suitcases and electrification, the coming together of diverse technologies was rejected by the very people who stood to benefit most from them. This is part of a deeper pattern. Many of us struggle with change, not because recombination is beyond our reach, but because we turn our backs on its possibilities. We assume that innovation is for creative types, or for techs in Silicon Valley. We unconsciously reject changes that might make our own jobs and lives more productive and fulfilled.

But there is one group of people who do not seem to be hampered by this barrier. People who are often behind the success stories we have mentioned, and who hold out lessons for all of us.

III

Take a look at the following list of names: Estée Lauder, Henry Ford, Elon Musk, Walt Disney, and Sergey Brin. Can you see what they have in common? On the surface, they look like a list of famous entrepreneurs, people who have made an impact on American society. But dig a little deeper and you will note that they share a pattern with dozens of others, including Dietrich Mateschitz, Arianna Huffington, and Peter Thiel, each of whom has helped shape the modern economy of the United States. The link between these people? They are all immigrants or the children of immigrants.

A study published in December 2017 revealed that 43 percent of companies in the Fortune 500 were founded or cofounded by immigrants or the children of immigrants, a number that rises to 57 percent in the top thirty-five companies. These companies produced $5.3 trillion in global revenue and employed 12.1 million workers worldwide in everything from tech and retail to finance and insurance. This is not an isolated finding. Immigrants make disproportionate contributions in technology, patent production, and academic science. A 2016 paper in the *Journal of Economic Perspectives* showed that U.S.-based researchers had been awarded 65 percent of Nobel Prizes over the preceding few decades. Who were these innovators? More than half were born abroad.

Different studies have shown that immigrants are twice as likely to become entrepreneurs. They account for 13 percent of the U.S. population, but 27.5 percent of those who start their own businesses. Another study, this time by Harvard Business School, showed that companies founded by immigrants grow faster and survive longer. Yet another showed that around a quarter of all tech and engineering companies started in the United States from 2006 to 2012 had at least one immigrant cofounder. This is not just about immigrants into the United States, it is a property of immigration more generally. Data from the 2012 Global Entrepreneurship Monitor shows that the vast majority of the sixty-nine countries surveyed reported higher entrepreneurial activity among immigrants than among natives, especially in high-growth ventures. None of these studies is conclusive on its own, but together they present a persuasive pattern.

Now, think back to the examples from the previous section. Why did existing luggage companies struggle to perceive the benefits of wheels? Why did established manufacturing companies struggle to fuse electrification with modified assembly lines? Why is it so often the people in the best position to gain from innovation who are blind to its opportunities? Could it be that when you are immersed within a paradigm, it is difficult to step beyond it? Think of luggage executives operating in 1950s America. Their lives were centered on conventional luggage. Their entire careers had been spent working with unwheeled suitcases. Their lives were bound up in the paradigm. It was a part of their worldview, their most basic frame of reference.

As for the executives and owners of the large industrial companies, they had worked with steam engines all their careers. This was their conceptual center of gravity, the way they filtered ideas and conceived of opportunities. It was the premise around which everything else orbited. This deep familiarity with the status quo made it psychologically difficult to deconstruct or disrupt it. As McAfee and Brynjolfsson write:

> It is exactly because incumbents are so proficient, knowledgeable, and caught up in the status quo that they are unable to see what's coming, and the unrealized potential and likely evolution of the new technology. . . . Existing processes, customers, and suppliers all blind incumbents to things that should be obvious, such as the possibilities of new technologies that depart greatly from the status quo.

In fact, this can be seen experimentally. A classic study by Robert Sternberg and Peter Frensch pitted experts and novices at the card game bridge. Unsurprisingly, the experts performed better. They were the experts, after all. But then the researchers made some structural changes to the rules. Instead of players who put out the highest card winning, this was reversed. This change had little effect on the performances of the novices. They rapidly absorbed the change and carried on. For the experts, who had much deeper familiarity with the rules and had spent years playing the game, the change was more disconcerting. They had much greater difficulty dealing with the disruption. Their performance declined.

This dovetails with our analysis of immigrants. They have experienced a different culture, a different way of doing things. When they see the business ideas in a new country, or a particular technology, they do not see something immutable. Irrevocable. Set in stone. They see something that could potentially be changed. Reformed. Amended, adapted, or subject to recombination. The very experience of seeing different places seems to offer psychological latitude to question conventions and assumptions. Let us call this the *outsider mindset.* Immigrants are not outsiders in the literal sense of physically standing outside a particular convention or paradigm. Rather, they are outsiders in the conceptual sense of being able to reframe the paradigm. To see it with fresh eyes. This provides them with the latitude to come up with rebel ideas.

Immigrants have another advantage, too, linked to the notion of recombination. Having lived with two cultures, they have greater experience bringing ideas together. They act as bridges, facilitators for "idea sex." If the outsider perspective confers the ability to question the status quo, diversity of experience helps provide the recombinant answers.

Years of patient empiricism have validated these truths. A study led by the economist Peter Vandor examined the capacity of students to come up with business ideas before and after a semester. Half the students went to live and study abroad during the semester, while the other half stayed in their home universities. Their ideas were then assessed by a venture capitalist. Those who studied abroad had ideas that were rated 17 percent higher than those who had not. Indeed, those who stayed in their home universities actually experienced a decline in the quality of their ideas over the course of the study.

In a different experiment, students were given a test of creative association. They were presented with sets of three words and asked to come up with a fourth word that links them together. One set was "manners, round, tennis." Can you think of the fourth word that links them? Another set was "playing, credit, report."[10]

Before the task, half the students were asked "to imagine living in a foreign country and, in particular, about the types of things that happen, how they feel and behave, and what they think during a particular day living abroad. They were then asked to think and write about this experience for several minutes." A control group was given a different task: they were asked to think about a day not living abroad but in their hometown.

What happened? Those who imagined living abroad were 75 percent more creative, solving more puzzles, and seeing connections that those who had focused on their hometown just couldn't see. Dozens of other experiments have found similar results, in multiple contexts. It is as if imagining living across national borders helps us step beyond conceptual borders.

This discussion isn't about travel, or even immigration. It is about the outsider mindset. After all, a fresh climate doesn't have to be geographical. Charles Darwin alternated between research in zoology, psychology, botany, and geology. This did not diminish his creative potential, but rather

---

[10] The answer to the first triad is "table" (i.e., table manners, roundtable, table tennis). For the second triad, the solution is "card" (i.e., playing card, credit card, report card).

enhanced it. Why? Because it gave him the chance to see his subject from the outside and to fuse ideas from diverse branches of science. One study found that the most consistently original scientists switched topics, on average, a remarkable forty-three times in their first hundred published papers.

A team at Michigan State University compared Nobel Prize–winning scientists with other scientists from the same era. The Nobel laureates were twice as likely to play a musical instrument, seven times more likely to draw, paint, or sculpt, twelve times more likely to write poetry, plays, or popular books, and twenty-two times as likely to perform as amateur actors, dancers, or magicians. Similar results were found for entrepreneurs and inventors.

Psychologists often talk about "conceptual distance." When we are immersed in a topic, we are surrounded by its baroque intricacies. It is very easy to stay there, making superficial alterations to its interior. We become prisoners of our paradigms. Stepping outside the walls, however, permits a new vantage point. Even if we have no new information, we have a new perspective. This is often considered to be a primary function of certain types of art. It is not about seeing something new, but about seeing something familiar in a new way. One thinks of the poetry of W. B. Yeats or the paintings and sculptures of Picasso. These great works create conceptual distance between the viewer of the painting and its object, the observer and the observed.

In a world where recombination is becoming the principal engine of growth, this could not be of greater significance. The growth of the future will be catalyzed by those who can transcend the categories we impose on the world, who have the mental flexibility to bridge between domains, who see the walls that we construct between disciplines and thought silos and regard them not as immutable but movable, even breakable.

This is why the outsider mindset is set to become even more powerful.

That is not to say that we don't need insider expertise; quite the reverse. We need both conceptual depth and conceptual distance. We need to be insiders and outsiders, conceptual natives and recombinant immigrants. We need to understand the status quo, but also to question it. We need to be strategically rebellious. To return to immigrants, there are doubtless additional reasons that explain their outsize contribution to innovation. The kinds of people who choose to migrate are likely to be comfortable with risk-taking. Given the barriers they often face, they are likely to develop resilience. But while these traits are important, they should not obscure the significance of being able to question the status quo and step beyond convention.

Catherine Wines, a British entrepreneur, puts the point well: "To become a visionary, you have to take the perspective of an outsider in order to see the things that are taken for granted by insiders. Possibilities and opportunities become most apparent when you are confronting a problem with a fresh perspective."

Wines founded a remittance company with Ismail Ahmed, a Somali immigrant, in 2010. Ahmed had arrived in London in the 1980s, having experienced firsthand the deep frustrations of receiving needed funds from family abroad, which were often delayed or lost. His early life, together with what he learned in his new home about digital solutions, led to a new venture: a company that allows users to send money home (or anywhere else) via text message. It is a classic example of recombination.

Bezos made the same point in his 2018 letter to shareholders. He talked about the importance of incremental innovation, doubling down on existing ideas, exploiting their value. Yet he also recognized that, if you want to innovate in more profound ways, you have to step outside your existing framework. His word for this captures the outsider mindset. He calls it *wandering*. Bezos says:

Sometimes (often actually) in business, you do know where you're going, and when you do, you can be efficient. Put in place a plan and execute. In contrast, wandering in business is not efficient . . . but it's also not random. It's guided—by hunch, gut, intuition, curiosity . . . [and] it's worth being a little messy and tangential to find our way there. Wandering is an essential counterbalance to efficiency. . . . The outsize discoveries—the "nonlinear" ones—are highly likely to require wandering.

Think about the implications for education. Labor experts predict that the children of today will have as many as a dozen jobs, the majority of which haven't yet been invented. In a fast-moving world, we will need to master not merely the art of invention but of personal reinvention. This is a world ripe for people who can question the status quo and who can transition beyond its boundaries, not least the ones we impose on ourselves. For if there is one paradigm in which we are deeply immersed, it is our own lives.

If it was difficult for luggage executives to question the status quo when it came to suitcases, how much more difficult is it to deviate from the script we are living every day? The default is integrated into our waking existence, our most basic frame of reference, the jobs we do, the skills we have, the lives we lead. Less salient are the skills we can yet build, and the opportunities we haven't yet considered. In short, we sometimes need to apply rebel ideas to our own lives.

Of course, sometimes it's great to have stability. To have continuity. But neither is there anything wrong with seizing opportunities, instead of inadvertently missing them, failing to grasp the equivalent of a wheeled suitcase or smart electrification idea in our own lives. What assumption-reversal techniques could I apply to what I am doing, and how I am doing it? Where is the potential for recombination?

Research led by Keith Stanovich of the University of Toronto measures one aspect of the outsider mindset. It is called the Actively Open-Minded Thinking scale. The questionnaire asks people if they agree or disagree with statements like "People should always take into consideration evidence that goes against their beliefs" and "A person should always consider new possibilities." Perhaps unsurprisingly people who score high on this scale are better at coming up with ideas, evaluating arguments, combating biases, and spotting fake news, even after controlling for cognitive ability.

There are many techniques that can help us transition to an outsider perspective, to see the familiar with new eyes, and to engage with new ideas. Michael Michalko, a former U.S. Army officer who has become a leader in creativity, advocates *assumption reversal*. You take the core notions in any subject or proposal, and simply turn them on their head. Suppose you are thinking of starting a restaurant. The first assumption might be "restaurants have menus." The reversal would be "restaurants have no menus." This provokes the idea of a chef informing each customer what he bought that day at market, allowing them to select a customized dish. The point is not that this will necessarily turn out to be a workable scheme, but that by disrupting conventional thought patterns, we arrive at new associations and ideas.

Think of how this technique might have altered the Industrial Revolution if at the time electrification had become available executives had reversed their defining assumption that "production processes are based on group power." Instead, they would have said that "production processes are *not* based on group power." Would this not have disturbed their assumptions, driven a new set of thoughts, and helped them escape their paradigm?

Or, to take a different example, suppose you are considering starting a new taxi company. The first assumption might be "taxi companies own cars." The reversal would be "taxi companies own no cars." Twenty years ago, that might have sounded crazy. Today, the largest taxi company that has ever existed doesn't own cars: Uber.

IV

Let's broaden the lens from individuals to institutions, to examine innovation from a wider perspective. What kinds of societies facilitate rebel combinations? Why are some places and epochs more creative than others? How does our analysis of diversity fit into the arc of history? Ideas, unlike physical goods, are not subject to diminishing returns. If you give someone your car, you cannot use it at the same time. If you come up with a new idea and share it with other people, however, its potential increases. This is known as *information spillover.* New people can bring their diverse perspectives to this new idea, improving it, augmenting it.

As Paul Romer, an economist who won the Nobel Prize for his work on innovation, put it, "The thing about ideas is that they naturally inspire new ones. This is why places that facilitate idea sharing tend to become more productive and innovative than those that don't. Because when ideas are shared, the possibilities do not add up. They multiply."

The key word here is "shared." Ideas can only spill over when people are connected with one another. Hero of Alexandria invented a steam engine in the first century AD, but news of the invention spread so slowly and to so few people that it may never have reached the ears of cart designers. Not only was the innovation lost on other people, but they lacked the opportunity to improve or recombine it, a point made by Ridley. Ridley also points to Ptolemy's astronomy, which was a great improvement on what went before (if not entirely accurate), but was never actually used for navigation because astronomers and sailors did not meet.

Innovation was isolated, deprived of cross-pollination, because people lived in structures—social, physical, moral—that lacked connectivity. There was no spillover.

When ideas are shared, they are not just transmitted to other minds, but they can now be combined with yet more ideas. Take the discovery of oxygen, which is typically credited to Joseph Priestley and Carl Wilhelm Scheele, as if they plucked the element from thin air. But to even start the search, they needed to know that air is made of distinct gases. This was not widely accepted until the second half of the eighteenth century. They also needed sophisticated scales to measure fine changes in weight, which didn't become available until a couple of decades earlier.

Priestley and Scheele were both creative, and had the outsider mindset willing to challenge the status quo, but they couldn't have made their breakthrough without being connected to a broader web of people and ideas. It was the diversity in their social network that enabled them to combine previously unconnected concepts, which then spilled back, inspiring new ideas and recombinations. This implies that our perspective on innovation should shift from one in which individuals are front and center to one in which new ideas and technologies emerge from a complex dance between individuals and the networks they inhabit.

In his book *The Sociology of Philosophies*, Randall Collins, a professor at the University of Pennsylvania, chronicles the intellectual development of pretty much every significant thinker in recorded history. He argues that the likes of Confucius, Plato, and Hume were, indeed, geniuses, but shows that their genius blossomed because they were situated at propitious nodes in the social network. Take a look, in figure 7, at Collins's attempt to recreate the network of Socrates, which reveals connections with virtually every major thinker.

For those interested in existentialism, figure 8 illustrates the network that encompasses Jean-Paul Sartre and Martin Heidegger.

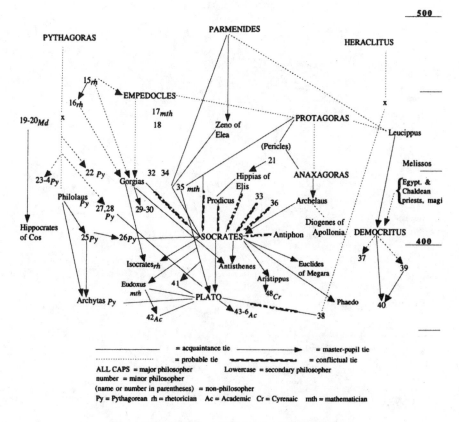

Figure 7. Centralization of the Greek network in Athens

Collins writes:

Intellectual creativity is concentrated in chains of personal contacts, passing emotional energy and cultural capital from generation to generation. This structure underwrites all manner of contexts: we see it in the chains of popular evangelists in Pure Land Buddhism, as among the masters of Zen, among Indian logicians and Japanese neo-Confucians. . . . The emotional energy of creativity is concentrated at the center of networks, in circles of

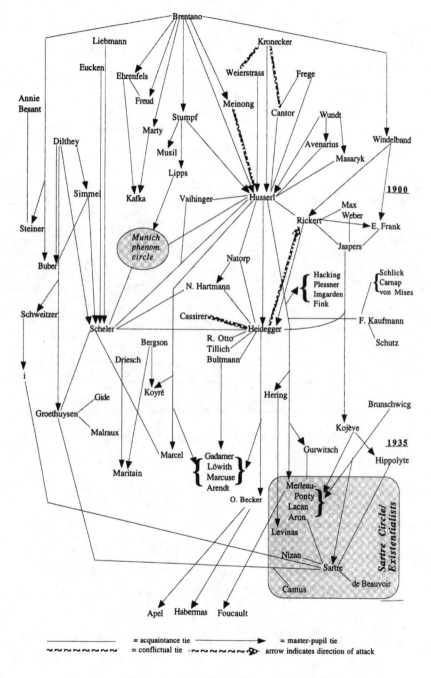

Figure 8. Network of phenomenologists and existentialists

persons encountering one another face to face. The hot periods of intellectual life, those tumultuous golden ages of simultaneous innovations, occur when several rival circles intersect at a few metropoles of intellectual attention and debate.

The social context of creativity offers a holistic perspective and an inspiring truth: that innovation is about the creativity of brains in a social network, but that the creativity of brains is also about the diversity of networks they are plugged into. The overall network of connected brains is what the evolutionary theorists Michael Muthukrishna and Joseph Henrich call the *collective brain*. They write:

A common perception of the source of innovation is Carlyle's "great man"—the thinker, the genius, the great inventor—whose cognitive abilities so far exceed the rest of the population, they take us to new places through singular, Herculean mental effort. They may stand on the shoulders of the greats of the past, but they see further because of their own individual insight; their own individual genius. We argue . . . that these individuals can be seen as products of collective brains; a nexus of previously isolated ideas.

This picture explains why innovations often occur in different minds at almost precisely the same time. For a long while, fate or providence was cited to explain why, across all of space and time, Darwin and Alfred Russel Wallace came up with versions of the theory of evolution virtually in the same month. Or why Gottfried Wilhelm Leibniz and Isaac Newton hit on the concept of calculus almost simultaneously. Fate started to seem like a rather unsatisfactory explanation, however, when historians realized that these "coincidences" are not the exception; they are the norm. As author Steven Johnson put it:

Sunspots were simultaneously discovered in 1611 by four scientists living in four different countries. The first electrical battery was invented separately by Dean Von Kleist and Cuneus of Leyden in 1745 and 1746. . . . The law of the conservation of energy was formulated separately four times in the late 1840s. The evolutionary importance of genetic mutation was proposed by S. Korschinsky in 1899 and then by Hugo de Vries in 1901, while the impact of X-Rays on mutation rates was independently discovered by two scholars in 1927. The telephone, telegraph, steam engine, photograph vacuum tube, radio—just about every essential technological advance of modern life has a multiple lurking somewhere in its origin story.

What is going on? How could these "independent" discoveries happen so often? They are the consequence of networked minds. When people are connected to similar people and ideas, they tend to make similar connections and discoveries.

We can see these truths at multiple scales simultaneously. Take a study by the anthropologists Michelle Kline and Rob Boyd on innovation rates in the Pacific islands. These islands are separated by hundreds of miles of water, making it possible to relate the speed of innovation to the size of the collective brain. The researchers found that the sophistication of the technology was strongly correlated with the size and interconnectedness of the population. Bigger networks permitted greater scope for recombination of ideas, competition between ideas, and information spillover.

Or take the state of Tasmania, an island 150 miles south of Victoria in Australia. When Europeans first came into contact with it in the late eighteenth century, the technology was astonishingly primitive; there are tribes dating from forty thousand years ago with a more sophisticated tool kit. The Tasmanians had one-piece spears, had reed rafts (which leaked), were unable to catch or eat fish (even though there were plentiful supplies), and drank from skulls.

How could they have become marooned with such basic technology? Joseph Henrich has pointed out that the puzzle fits into place when you realize that twelve thousand years ago, sea levels rose, flooding the Bass Strait, thus cutting Tasmania off from the rest of Australia. For more than twelve millennia, they were disconnected from the broader network of ideas, shrinking the collective brain.

A tiny population was now isolated, with the risk that a skilled craftsman might die before he had taught his apprentices, leading to the disappearance of hard-won innovations. More significantly, they could no longer communicate with Australia: couldn't learn, improve, recombine. As the Pama-Nyungan expansion gained pace across the strait, Tasmania—which had the same technology at the time of the flooding—went into precipitous decline.

You can make the same point by comparing technologies. Each year, Henrich shows new students unlabeled tool kits from four different populations: eighteenth-century Tasmanians, seventeenth-century Australian Aborigines, Neanderthals, and humans from thirty thousand years ago. When the students are asked to assess the cognitive abilities of the toolmakers, they *always* give the same reply. They rate the Tasmanians and Neanderthals as lower in cognitive ability than the Aborigines and humans from thirty thousand years ago because the tools are less sophisticated.

But this is wrong. Why? Because it is impossible to determine innate individual cognitive abilities from the complexity of tool kits. The reason is that innovation is about not just individuals but connections. Think back to the Tasmanians before and after the flooding. The populations were genetically identical but the relative strength of their tool kits could not have been more different.

LUGGAGE EXECUTIVES FAILED to adopt wheeled suitcases because they were stuck in the status quo. The opportunities of recombination were missed because of the conceptual walls imposed by their paradigm. They were held back by an insider mindset.

The same analysis applies to the bigger picture. Tasmania struggled to innovate because it was separated from the possibilities of recombination—not because of an insider mindset, but because of a flood. They were physically rather than psychologically separated from new ideas. The severance in the network structure put literal constraints on innovation.

This kind of separation can be ideological, too. For many centuries women were excluded from the network of ideas. An entire social group faced a barrier brought about not by flooding, but by bigotry. This continued into the Enlightenment. As Carol Tavris, the social psychologist, writes, the Enlightenment "narrowed the rights of women, who were . . . barred from higher education and professional training." This was socially unjust for women—*but it also dramatically reduced the creativity of men.* By severing males from the insights that could have been brought by half the population—the diverse perspectives, information, context, and discoveries—the collective brain was serially diminished. Whatever else we may say about the pace of innovation over the course of human history, it would have been far faster if the idea network had included women.

These points can be made using simple math. Henrich invites us to imagine two tribes seeking to invent a particular technology: say, a bow and arrow. He also asks us to imagine that these tribes have different attributes. The geniuses are smart. They have huge brains. The networkers, on the other hand, are sociable. They like to interact. Now, suppose that a genius is so smart that, through individual effort and imagination alone, he will create the innovation once every ten lifetimes. A networker, on the other hand, will only create the innovation once in every thousand lifetimes. We might say, then, that the geniuses are a hundred times smarter than the networkers.

However, the geniuses are not very social. They only have one friend in their network they can learn from. The networkers, on the other hand, have ten friends, making them ten times more social. Now, after everyone has tried to invent the bow and arrow for themselves, and then tried to learn from their friends, from whom they have, say, a 50 percent chance of learning in each encounter, in which population will the innovation be more common?

The answer is counterintuitive. Among the geniuses, only 18 percent of people will have the innovation. Half of those will have figured it out on their own. Among the networkers, on the other hand, 99.9 percent will have the innovation. Only 0.1 percent will have solved it on their own, but the rest will have learned it from friends. And each of these will now have an opportunity to improve on the innovation, transmitting insights back into their networks. The result is clear—and is corroborated by field data, lab experiments, and dozens of historical examples. As Henrich puts it, "If you want to have cool technology, it is better to be social than smart."

V

Route 128 is a highway that starts at Norfolk County in the south and winds up past the western suburbs of Boston, before culminating, right on the coast, by Gloucester, a fishing town that was the setting for Rudyard Kipling's famous book *Captains Courageous*.

When Jonathan Richman wrote "Roadrunner," his song about Route 128, listed by *Rolling Stone* magazine as one of the greatest five hundred songs ever recorded, the highway encompassed what many believed was an enduring economic miracle. In 1975, the technology complex employed tens of thousands of workers and boasted six of the ten largest tech companies in the world. Wang Laboratories, Prime, and Data General were giants of the sector. At its height, Digital Equipment Corporation (DEC) boasted a workforce of 140,000, the second-biggest employer in the state. The western section of the route was dubbed America's Technology Highway. *Time* magazine called it the Massachusetts Miracle.

The Santa Clara Valley, on the other hand, was an agricultural region over three thousand miles away in California, largely devoted to apricot farming. The apricots were juicy and fragrant, but they were rather a long way from chips and semiconductors. Most of the local industry was based on small-scale food processing and distribution. As one historian said, "There wasn't much going on." The Santa Clara region started to change in 1956 when William Shockley, a physicist and inventor, moved from an unsuccessful stint at Raytheon, a Massachusetts firm with an interest in transistors, to Mountain View, a small town toward the southern end of the San Francisco Peninsula. Over time, this led to a growing concentration of firms in the valley, including Fairchild Semiconductor International.

By the 1970s, the Santa Clara Valley had spawned a sobriquet of its own—Silicon Valley—but it remained very much in the shadow of the Massachusetts Miracle. The Boston firms had the classic economic advantages. Land and office space costs were significantly lower, as were the wages and salaries of workers, engineers, and managers. There were other differences, too. The Boston firms were buttoned-up. They wore jackets and ties. The Silicon Valley rebels were more laid-back, preferring jeans and T-shirts. They had different ways of talking, and different terminology. Yet, these were the superficial differences. The key differences consisted in the structure of networks, and the dynamics of information spillover. And these would prove utterly decisive.

The Route 128 firms had scale. They made chips and boards, monitors and frames, all internally. They even made disk drives. This vertical integration made sense economically. It meant that they had impressive efficiency in production. But this integration had another, less remarked-on consequence (a consequence it didn't need to have). These large firms became socially isolated. Gordon Bell, a vice president of DEC, said, "DEC was a large entity that operated as an island in the regional economy." Glenn Rifkin and George Harrar, biographers of Ken Olsen, a cofounder of DEC, described the company as "a sociological unit, a world unto itself." AnnaLee Saxenian, a sociologist who covered the tech wars in her classic book *Regional Advantage,* writes, "Route 128's enterprises adopted autarkic practices."

As the firms became isolated, they also became fiercely proprietorial. Wang hired private detectives to protect its ideas and property. People socialized only with people from their own firm. There were precious few forums or conferences that brought engineers together. "Practices of secrecy govern relations between firms and their customers, suppliers and competitors," Saxenian writes. Another said: "The walls got thicker and thicker, and higher and higher."

The desire for secrecy made sense, in its own terms. The executives didn't want other firms stealing their ideas. But it also represented a trade-off, unspoken but profound. By severing their engineers from the broader network, they inadvertently stifled the interplay of diverse insights, fusing, recombining, jumping forward in unpredictable ways: the complex dance of innovation. Route 128, then, was characterized by what network theorists call a *vertical dynamic.* Ideas flowed within these hierarchical organizations, but not outward. "Information on technologies remained trapped within the boundaries of individual corporations, rather than diffusing to other local firms and entrepreneurs," Saxenian writes. "There was very little horizontal transmission."

You could even sense the social isolation in the physical terrain along the highway, the way that the local companies distanced themselves from each other. "The technology companies were scattered widely along the corridor and increasingly along the outer band . . . with miles of forest, lakes, and highway separating them. The Route 128 region was so expansive that DEC began to use helicopters to link its widely dispersed facilities."

On the surface, at least, Silicon Valley seemed less suited to the high-tech sector. The region didn't enjoy tax benefits to help them catch up with Route 128, nor did they have additional state support in, say, defense spending. And, as already noted, costs were higher in land, office space, and wages. And yet Silicon Valley had something more powerful, an ingredient that rarely finds its way into conventional economic textbooks. You get a sense of this ingredient by reading Tom Wolfe in a famous essay on the Valley:

> Every year there was some place, the Wagon Wheel, Chez
> Yvonne, Rickey's, the Roundhouse, where members of
> this esoteric fraternity, the young men and women of the
> semiconductor industry, would head after work to have a drink
> and gossip and brag and trade war stories about phase jitters,
> phantom circuits, bubble memories, pulse trains, bounceless
> contacts, burst modes, leapfrog tests, p-n junctions, sleeping-
> sickness modes, slow-death episodes, RAMs, NAKs, MOSes,
> PCMs, PROMs, PROM blowers, PROM burners, PROM blasters,
> and teramagnitudes, meaning multiples of a million millions.

In Silicon Valley, people socialized, ideas fizzed around, giving them a chance to meet and mate, to recombine, and to trigger yet new ideas. "There is a velocity of information here that is very high," one observer said. "The region's dense social networks and open labor markets encourage experimentation and entrepreneurship," Saxenian writes. "The standing joke was that if you couldn't figure out your process problems, go down to the Wagon Wheel and ask somebody."

This is what is sometimes called *horizontal information flow*: the kind that travels from engineer to engineer, firm to firm, spilling over all the time. Information circulated not merely within institutions, but between institutions. The process was bolstered by the geographical topology: unlike the widely scattered islands of Route 128, the firms in the Valley "clustered in close proximity to one another in a dense industrial concentration." Spaces like the Wagon Wheel, neutral ground between the firms, were hubs of recombination, cauldrons bubbling with people with different perspectives and paradigms. Insiders on one topic or technology were outsiders on another and vice versa, creating vast diversity of thought.

One such space was the Homebrew Computer Club, started by enthusiastic hobbyists, who held their first meeting in a garage. The logic was contained in the very first newsletter, posted in March 1975. "Are you building your own computer? Terminal? TV Typewriter? I/O device? Or some other digital black-magic box? If so, you might like to come to a gathering of people with like-minded interests. Exchange information, swap ideas, talk shop, help work on a project, whatever." (That first meeting took place just a few blocks from where, a few decades earlier, two men named Bill Hewlett and David Packard started experimenting with electronic equipment in a different garage.)

The inaugural meeting set the stage for what would follow. Ideas bubbled up as if from a shaken soda can. There were only a few hundred personal computers in existence at the time, but as the conversation fizzed, the assembled group came up with dozens of ideas for potential home use: text editing, storage, games, educational uses. One person even suggested using a computer system to control home functions such as the alarm, heating, and sprinkler systems.

One of the attendees at that meeting was a bearded enthusiast in his midtwenties. Shy and soft-spoken, he listened as the discussion danced around the terrain of personal computing. He had built his own processors, played around with chips, but he was now surrounded by a conversation that was the sociological equivalent of being plugged into thirty new brains, each with its own insights, diverse perspectives, specialist information, and rebel ideas.

As they discussed the Altair 8800, the first personal computer, which was sold to hobbyists in a build-it-yourself kit, he was intrigued. He had never seen one before. Then a data sheet from the 8800 was handed out, which fired up his mind. "It was a meeting that grabbed my attention for life," he later said. "It was a Eureka moment for me. . . . I took this data sheet home and was shocked to find that the microprocessor had gotten to the point of being a complete processor of the type I'd designed over and over in high school. That night the full image of the Apple I popped in my head."

The hobbyist's name was Steve Wozniak. Thirteen months later, he would start the Apple Corporation from the ideas that fused in his head that night (could there be a more exquisite example of a rebel combination?). His cofounder was another attendee at the Homebrew: Steve Jobs.

It is symptomatic that forums for idea exchange—whether restaurants, cafés, or organically created clubs—were conspicuously absent along Route 128. There was *no demand*. Jeffrey Kalb, who worked in Massachusetts in microcomputing before moving to Silicon Valley, said, "I was not aware of similar meeting spots in Route 128. There may have been a lunch spot in Hudson or Marlboro, but there was nothing of the magnitude of Silicon Valley hangouts." Route 128 companies didn't neglect these things as an act of deliberate self-sabotage. They were creative and smart, but had not made an essential conceptual leap. Innovation is not just about creativity, it is also about connections. They were somewhat like the geniuses in the thought experiment. They had originality, but lacked sociality. Diversity existed but was not exploited. The companies, like Tasmania, were islands separated by high waters. As Saxenian writes, "The networking and collaborative practices that typified Silicon Valley never became part of the mainstream business culture of Route 128, and the region's new management models only partially departed from traditional corporate practices."

In 1957, fifteen years before "Roadrunner" was penned, Route 128 employed more than twice the number of workers in the tech sector than Silicon Valley. They had established companies such as Sylvania, Clevite, CBS-Hytron, and Raytheon. These firms accounted for a third of the entire nation's transmitting and special-purpose receiving tubes and a quarter of all solid-state devices. In 1987, fifteen years after the song was written, this gap was reversed with elegant symmetry. Now, Silicon Valley employed two times as many tech workers as Route 128. By 2000, the isolated corporate islands of Boston had all but disappeared, somewhat like the technology on Tasmania.

We should note that competition between firms (even insular firms) is a form of information discovery at the level of the system. When institutions go head-to-head, we find out which ideas work and which do not. Companies with poor ideas go bankrupt, the successful firms are copied, and the system adapts. Well-functioning markets are a powerful engine of growth and contribute to the expansion of the collective brain. What the analysis of this chapter reveals, however, is the danger when information gets trapped within institutional boundaries. This is bad both for the system, because it evolves more slowly, and for the institutions themselves, which struggle to innovate.

Consider, too, that the fissures in the network along Route 128 both contributed to insularity in the incumbent companies and were exacerbated by it. This was a dangerous synergy. The more people retreated into their silos, the more they perceived new ideas not as opportunities but as threats.

Mitch Kapor, the founder of Lotus Development Corporation in Silicon Valley, talks of a "bizarre" meeting with Olsen, during which the latter seemed incapable of grasping the significance of personal computers. Kapor said:

> Some of my most sobering moments . . . were seeing how those
> guys weren't getting it, and were dooming themselves. Olsen had
> himself designed the case for the DEC personal computer, and he
> was banging on it and showing me how solid it was. I was going,
> "What planet am I on? This has nothing to do [with it]." But in
> his world, when computers were on factory floors and so on, they
> needed to be robust. That really mattered. This didn't matter
> at all.

As for Silicon Valley, the region was forging ahead, insiders and outsiders colliding, diverse concepts recombining. The net result was a maelstrom of high-velocity information flow. As Larry Jordan, an executive at Integrated Device Technology, said in a seminal interview in 1990, "There is a unique atmosphere here that continually revitalizes itself by virtue of the fact that today's collective understandings are informed by yesterday's frustrations and modified by tomorrow's recombinations. . . . Learning occurs through these recombinations. No other geographical region creates recombination so effectively with so little disruption. The entire industrial fabric is strengthened by this process."

VI

The science of recombination presents us with a compelling vision. Innovation is about breaking down walls. Some walls are good, of course. Most of us value privacy. Most companies need to protect intellectual property. Most institutions need specialists who, in turn, need space to do their job. But we often get the balance wrong, leaning toward insularity, not because we don't value the insights of people who think differently from ourselves, but because we underestimate their value. This is another aspect of homophily. We are comfortable in our own silos, our own categories, our own conceptual milieus.

This is true even of science itself. Too often, academics speak to academics in the same discipline. This is fine, but only to a point. When historians only talk to historians, and economists to economists, they undermine their capacity to understand the very phenomena they seek to explain. Much of this book is taken from the work of academics with bracing outsider perspectives, working in multidisciplinary groups, replete with diversity in gender and ethnicity: academics who are enriching our comprehension of the world.

Some of these thinkers struggle to get published in scientific journals. The reason is that parts of academia have become like conceptual islands, self-referential peer-reviewing groups who struggle with anything outside the paradigm. There is also an insufficient recognition that much of the greatest science is recombinant. Successful scientists are not just those with deep knowledge of their own terrain, but those who have the imagination to peer out into the broader constellation, looking for meaningful cross-pollination. This is how they discover rebel combinations.

Given what we have learned, it is perhaps unsurprising that network theory is moving center stage in multiple fields. The math was formalized by Leonhard Euler in the eighteenth century, but the basic ideas are intuitive and practical to apply. In building design, architects are now curating spaces that maximize the scope for connections. Instead of closed-off cubicles and walled offices, the idea is to bring people away from their desks, to create areas where people feel encouraged to mingle, make chance encounters, and engage with outsider perspectives.

One leader who grasped these truths intuitively was Steve Jobs. When he was designing the building for Pixar, the animation company he bought from George Lucas in 1986, he made the decision to create just one set of toilets. These were in the atrium, meaning that people had to traipse across from all over the building. It seemed inefficient, but it forced people out of their usual niches, and led to a symphony of chance encounters. "Everybody has to run into each other," Jobs said.

Or take Building 20 at MIT. This was not a beautiful structure. According to one essay on the building, it was "hastily constructed of plywood. It leaked. It had bad acoustics and was poorly lit, inadequately ventilated, very confusing to navigate (even for people who had been working there for years), and scorching in the summer and freezing in the winter." And yet it nevertheless spawned astonishing innovation, including the construction of the world's first atomic clock and the development of modern linguistics, one of the earliest atomic particle accelerators, stop-action photography, and more. Jerome Lettvin, the cognitive scientist, called it "the womb of the institute."

Why was the building so conducive to innovation? The lack of a formal structure caused scientists from different subject areas to bump into each other. Amar Bose, for example, started hanging out in the acoustics lab while taking a break from his dissertation, as the lab was just down the hall from his own department. He would later invent an innovative, wedge-shaped speaker (and establish Bose Corporation).

Inhabitants of the building over the years included the Acoustics Lab,

Adhesives Lab, Linguists Department, Guided Missiles Program Office, Lab for Lighting Design, Office of Naval Research, Model Railroad Club, and more. "At this time in history, these researchers would have never shared the same facilities—the biologists would have studied in the life science building, and the lighting designers would have been drafting away in the architecture building," writes David Shaffer, an architect. "Scientists from diverse fields got to know one another in a unique and exciting way that produced unparalleled cross-departmental collaboration."

Another unusual feature of Building 20 is that the rather flimsy walls could be torn down when they got in the way of fruitful collaboration. "If you want to wire from one room to another, you don't call Physical Plant [maintenance]," Paul Penfield, the engineering professor, has said. "Instead you get out a power drill and jam it through the wall." In his book *Messy*, Tim Harford writes, "Who would have thought that throwing the electrical engineers in with the model railway club would result in hacking and video games?"

You can see the power of networks in the history of cultural institutions, too. Soccer has been an incubator of recombination, particularly in the domain of tactics. This can be seen in everything from the WM approach of the legendary Spurs manager Herbert Chapman to the defensive Italian *catenaccio*. The economist Raffaele Trequattrini has shown that these innovations led to a sustained competitive advantage.

The same can be said of the soccer revolution in Holland, perhaps the most vivid instance of recombination in sports. Soccer fans might be surprised to discover that Dutch soccer was once highly insular. Ideas from beyond the game were seen as threats, not opportunities. When, in 1959, a new physical therapist arrived at Ajax and saw that the medical facility consisted of a wooden table and blanket, he offered to buy a modern treatment table. "Don't poison the atmosphere," the coach responded. "We've been doing it for fifty years on this table."

It took a young coach with an outsider mindset called Rinus Michels to challenge this insularity. He imported ideas from beyond the game to not only transform tactics and training, but inspire professionalization. Before his time, almost all the players worked in normal jobs away from the game, including Johan Cruyff, one of the greats of Dutch soccer, who worked shifts at a local printing factory. Training became more "imaginative, intensive, and far more intelligently focused."

In his fine book *Brilliant Orange*, David Winner traces these shifts to wider trends in Dutch society which was, itself, opening up to new ideas. "After 20 years of peace, there were unparalleled opportunities for international cultural cross-pollination. . . . Nowhere was youth rebellion fueled by so surreal, anarchic, and theatrical a sense of playfulness as in Amsterdam."

Cruyff was central to the transformation. Karel Gabler, a former youth coach who "grew up amid the ruins of Amsterdam's old Jewish district, where the 1960s seemed like an eruption of color into a world of monochrome," said, "Cruyff got into all kinds of conflicts because he started asking the question the whole generation was asking: 'Why are things organized like this?'"

In his book *The Talent Lab*, the journalist Owen Slot examines the success of British Olympic sports, which shifted from a low point of one gold medal at the Atlanta Games of 1996, to twenty-nine at the London Games of 2012. One of British sports' key appointments was Scott Drawer as head of research and development, a scientist with a PhD in sports science and a thirst for new ideas. One of his first actions was to look beyond sports, to academia and industry, to find engineers and inventors who might bring novel insights to the question of how to help athletes go faster. His group came together at the Wagon Wheel of its time, a meeting room in Sheffield, where individual scientists were plugged into a new, more diverse network. Drawer has said:

> It wasn't necessarily the best [group] in terms of academic expertise, but it was the best in terms of creativity, people who would listen, be curious, want to explore. The naiveté was a real strength. . . . When you get people in a place with a good environment where people can think like that, you can be amazed where it goes.

The recombinant results were remarkable:

> F1 technology to help build winter Olympic bobsleighs, British Aerospace technology to build the skeleton sled on which Amy Williams won gold in Vancouver . . . sensors for swimmers to perfect the tumble turn . . . "hot pants" for cyclists to wear to keep their muscles warm between races, liquid repellent that coated the spray skirt of Ed McKeever's canoe in which he won gold in London.[11]

You can see the same pattern throughout history. Epochs that have managed to tear down the barriers between people, which have facilitated meaningful interaction, have driven innovation. There are dozens of examples, but one of the most remarkable was eighteenth-century Scotland, which despite being something of a backwater for centuries and having

---

[11] Tim Wigmore, a British sports journalist, argues that many technical innovations in sports are, when you take a closer look, recombinant in nature. Indian cricketers refined the reverse sweep by drawing on insights from tennis. Novak Djokovic learned his famous slide by incorporating ideas from his love of skiing. The same pattern applies to the "flop" of high jumper Dick Fosbury, the "tomahawk" serve of female table tennis player Ding Ning, and the unusual eye-tracking techniques of rugby player Danny Cipriani.

just endured a century of political turmoil, emerged as a hub of the Enlightenment.

Scotland had developed an unusually extensive network of parish schools throughout the lowlands by the beginning of the eighteenth century and had five universities (University of St. Andrews, University of Glasgow, University of Edinburgh, Marischal College, and University of Aberdeen) compared with just two in England. All these institutions had chairs in mathematics and offered high-quality lecture-based education in economics and science.

The scene in Scotland was also highly social: "It was not a business of isolated individuals working in country estates, or of secluded academics, cloistered within unworldly universities. The scene was convivial." Academics, scientists, and merchants mixed together, not least at the lattice of clubs and societies that sprang up around this time. As one scholar put it, "The interconnections and cross-fertilization between disciplines . . . is one of the remarkable features of the Scottish scene. Geologists associated with historians, economists with chemists, philosophers with surgeons, lawyers with farmers, church ministers with architects."

The Oyster Club had among its founders the economist Adam Smith, the chemist Joseph Black, and the geologist James Hutton. The Select Society included the architect James Adam, the medic Francis Home, and the philosopher David Hume. These were the Wagon Wheels and Roundhouses of the Scottish Enlightenment, ideas colliding and diffusing.

The blossoming of knowledge was remarkable. Hume wrote masterpieces in moral philosophy, political economy, metaphysics, and history. Adam Smith, who was close friends with Hume, penned *The Wealth of Nations*, which remains arguably the most influential work in the history of economics. James Boswell wrote *An Account of Corsica*, James Burnett founded modern comparative historical linguistics, and Hutton was a pioneering geologist. Sir John Leslie conducted important experiments with heat while Joseph Black discovered carbon dioxide.

Consider the thinkers of the Scottish Enlightenment in isolation, and you might conclude that the nation was blessed with an unusual number of great minds. But take a step back and we see the fertile collective brain that gave those minds the chance to blossom. As one visitor put it, "Here I stand, at what is called the Cross of Edinburgh. and within a few minutes take fifty men of genius by the hand."

ECHO
CHAMBERS

I

Derek Black was still in primary school when he declared his commitment to white supremacy. By his teens he was helping manage Stormfront, a forum that started out as an online bulletin board, but soon became the web's first hate site. A 2001 *USA Today* article called Stormfront "the most visited white supremacist site on the net." Black posted regularly and moderated comments, helping white nationalists build an online community. He was committed and strategic and soon became central to both the website and the broader movement. Quick-witted and articulate, he was regarded by many as the crown prince of the cause, the person to whom people looked for new ideas and slogans.

By his late teens, Derek was given his own slot—*The Derek Black Show*—on AM radio. He advocated the writings of Ernst Zündel, a German publisher who promoted Holocaust denial, and interviewed supremacist leaders such as Jared Taylor and Gordon Baum. The show was so popular that he was given a daily program. He was a natural broadcaster.

He continued to manage and promote Stormfront, seeking to play down the link between users of the forum and episodes of mass violence. One study showed that in the five years leading up to 2014, Stormfront members murdered nearly a hundred people, with seventy-seven murders perpetrated by Anders Behring Breivik, the man behind the 2011 Norway attacks—a murder rate that "began to accelerate rapidly in early 2009, after Barack Obama became the nation's first Black president."

As a young adult, Derek became a fixture at supremacist gatherings, electrifying audiences as a platform speaker. He was developing a reputation as a potent thinker on racial repatriation. He won a seat on the Palm Beach Republican Executive Committee, but he was denied the position when the party found out about his extremist views.

On the night Barack Obama won the 2008 presidential election, Stormfront crashed due to the sheer volume of traffic. This was a heady time for white nationalism, with their ranks swelling daily and the new president receiving more than thirty death threats a month. Not long afterward, Black was given high billing at a meeting of a white-rights conference in Memphis. In his scintillating book *Rising Out of Hatred*, the journalist Eli Saslow sets the scene:

The Klansmen and neo-Nazis arrived for their meeting in the fall of 2008 dressed in suits with aliases written on their name tags and began sneaking into the hotel just after dawn. They walked past the protesters waving rainbow flags on the sidewalk, past the extra state troopers stationed outside the hotel lobby, past the FBI informants hoping to infiltrate their way inside. . . . One suburb declared a state of emergency so it could hire additional police officers; another issued a temporary ban on all public gatherings. But by 7:00 on Saturday morning, about 150 of the world's preeminent white nationalists had gathered inside a nondescript hotel conference room where a small sign hung on the wall. "The Fight to restore White America begins now," it read.

In some ways, Derek was born to be a white nationalist leader. Don, his father, had joined the Ku Klux Klan in college and was rapidly promoted to Grand Wizard. In 1981, he was arrested alongside other white supremacists carrying dynamite, tear gas, and other materials en route to attempt a coup on the island of Dominica. "They hoped to turn it into a white utopia," Saslow writes. He was sentenced to three years, learning computer skills during his incarceration that would enable him to set up the Stormfront website—of which his son would later become such an indispensable part.

As he witnessed Derek's meteoric rise, Don was filled with pride. "I never thought it would feel so good to play second fiddle in my own house," he said. He felt that Derek had many strengths that he himself lacked, not least a supple intellect. The youngster was able to coin phrases that captured the public imagination. When Derek talked of "white genocide" caused by mass immigration, Don noticed the way that it took hold, filtering into the mainstream.

Chloe, Derek's mother, also had a long association with the white nationalist movement. She married David Duke, one of the highest-profile members of the Ku Klux Klan, in her twenties and had two daughters with him. A few years after their divorce, she married Don, whom she had known for years in the circles of white nationalism. Duke was the best man at their wedding.

Duke, the de facto leader of white supremacism in the United States, had spent his life trying to bring white ideology into the political mainstream. When he ran for governor of Louisiana in 1991, he won a majority of the white vote but narrowly lost the election. He was Derek's godfather and "like a second dad." He spent Christmases with the Blacks, socialized with them, and nurtured young Derek. It felt like he was grooming his successor.

By his late teens, Black was as well versed in the doctrines of white nationalism as he was comfortable in his own skin. His red hair fell down to his shoulders. He wore a black cowboy hat. He was personable and charming. People liked him. He didn't use racist slurs or advocate physical attacks, instead using softer language to articulate his ideology. He wanted America to be all-white and for minorities, ultimately, to be forcibly expelled.

At the meeting in Memphis, Duke's excitement was palpable as he presented the young prodigy to the mass of assembled supremacists. It felt like a defining moment. "The future of our movement is to become fully mainstream," Duke said. "I'd like to introduce you to the leading light of our movement. I don't know anyone who has better gifts. He may have a much more extensive national and international career than I've had. . . . Ladies and gentlemen, here is Derek Black."

II

The University of Kansas is the largest university in the Sunflower State. Founded in 1865 on a hill in the town of Lawrence, it has since expanded to five campuses, mainly in woodland, and is regarded as one of the most beautiful academic settings in the United States. "We embrace our role as the state's flagship university and a premier research institution, serving the state, nation, and world," the website says. "We celebrate the energy and compassion that infuse the Jayhawk spirit."

Speak to the students and academics, and you get a sense of not just the social buzz of the university, but also its scale. In total, there are almost thirty thousand students, hailing from all corners of the United States, as well as from around the world. There are almost three thousand students who are nonwhite almost six thousand are drawn from beyond the state of Kansas, and almost two thousand are over the age of twenty-five. This is a diverse population.

Stand within the perimeter of any university, whether in Kansas or anywhere else, and you gain a fleeting impression of the organic way in which social networks emerge from the broader collection of students. People gravitate toward each other after lectures to hang out in the bars and clubs, developing friendships, many of which will have lifelong duration. Most people stay in touch with university friends long after graduation.

Over recent years, the way in which social networks are formed has become a major focus of scientific investigation. There have been many studies, but one of the most fascinating was conducted by Angela Bahns, an American psychology professor, who looked at academic institutions in the state of Kansas. One of the establishments they analyzed was the University of Kansas. The researchers observed the students, watched as they hung out with their friends, and then gave them questionnaires so that they could probe the way they built friendships and social groups. In addition to the University of Kansas, they also studied five smaller universities in the state: Baker University in Baldwin City, Bethany College in Lindsborg, Bethel College in North Newton, Central Christian College in McPherson, and McPherson College, also in McPherson.

When I say these other universities are smaller, I mean much smaller. Baker University, a wonderful college with a rich history (it is the oldest academic establishment in the state, founded in 1858), has only three residence halls and two apartment buildings for students. It has a strong academic reputation and offers a selection of courses but cannot match the scale of facilities of the marquee institutions in the United States.

Whereas the University of Kansas had a student population of more than 27,000 in 2009, the other five universities had an average of barely 1,000. McPherson College had 629 students, Bethany College had 592 students, and Bethel College had just 437 students. These colleges also, by implication, have less overall demographic diversity. Bethel has just 105 students from outside Kansas, while Baker University and McPherson College have no students at all from overseas.

Bahns wanted to know how these differences in background conditions influence the characteristics of the social networks within the institutions. How would they shape the way people made connections with each other? How would they influence the type of people the students socialized with and how they built long-term friendships? The answer, at an intuitive level, seems obvious. The University of Kansas, by sheer dint of scale, affords far more in the way of opportunities to meet people who think in different ways, who come from different backgrounds, who have different perspectives. The university is a cosmopolitan institution by virtue of its more formidable reach.

Bethel College, on the other hand, might be an impressive institution, but its small size implies far more restricted opportunities for diverse people to mix. For all its benefits of intimacy and warmth, a tiny student population must surely curtail the possibilities for meaningful interaction with people who think differently, act differently, or just look a bit different.

When Bahns looked at the data, however, she found the complete opposite. The social networks at the University of Kansas were more homogenous, not just in terms of attitudes and beliefs, but also politics, moral convictions, and prejudices. "It was a clear result, and completely different to what most people expect," Bahns told me. "When people are a part of broader communities, they are likely to construct networks that are narrower."

How is this possible?

Think back to the two campuses. At Kansas, there are lots of people. They are diverse, to be sure, but diversity has a paradoxical property. It means there are a lot of different people to potentially interact with, but it also means that there are many people who are very much like oneself. If one wants to hang out with the like-minded, they are not so very difficult to find. Sociologists call this *fine-grain assorting.*

At a smaller college with fewer people, on the other hand, there is less overall diversity. But this means that it is almost impossible to find someone who thinks or looks exactly like oneself. You have to compromise, to accept some minimum level of difference. The smaller the overall amount of diversity in the background population, the greater the limitations on finding conformity. Bahns says:

> It sounds ironic, but it is quite predictable. In the smaller
> universities, there are fewer available choices, and people have
> to make connections with people who are comparatively more
> different. When the campus size is big, on the other hand,
> there is a greater opportunity for students to "fine-tune" their
> social network. They can pursue people who are very similar to
> themselves.

Bahns's experiment has parallels with other studies, in many parts of the world, and in different contexts. In one experiment led by Paul Ingram, a professor at Columbia Business School, one hundred business people were invited to an after-work mixing event in New York. It took place at 7 p.m. on a Friday evening at a reception hall of the university, and the researchers could not have done more to encourage intermingling. In the center of the room was a large table of hors d'oeuvres, by one wall there was a table with pizza, and by another wall there was a bar serving beer, wine, and soft drinks.

The participants, on average, knew about a third of the people in the room, but were unknown to the majority. This, then, was an opportunity to broaden their social network, to connect with diverse people. Indeed, many of the attendees specifically said in a pre-mixer survey that their main purpose in attending (alongside winding down) was to make new contacts. All attendees were fitted with electronic tags, which could not hear what was being said, but which could track the encounters, along with their duration. This enabled the researchers to "build a dynamic network that captured encounters throughout the event."

What happened? Whom did the attendees end up talking to? Did they seek out new people and expand their networks, as per their stated objective? In fact, the opposite happened. As the researchers put it, "Do people mix at mixers? The answer is no—or not as much as they might. . . . Our results show that guests at a mixer tend to spend the time talking to the few other guests whom they already know well."

THE MOST POWERFUL constraint on the growth of the collective brain in our species' early history was social isolation. Nomadic groups of hunter-gatherers were often geographically dispersed and had few means of communication. When groups started living closer together following the agricultural revolution, sociality was constrained by the many barriers that can exist between human groups, both physical and psychological. We noted that Tasmania went backward when it was separated from the broader Australian ecology.

Today, however, we live in a radically different era. People are connected not just socially but digitally. The Internet has created a hyperspace that spans the globe and can be triggered instantly. We have unprecedented access to diverse opinions, beliefs, ideas, and technologies, all at the click of a mouse. This was, of course, the original vision of the Internet by Tim Berners-Lee: a place where a scientific community could share research and ideas. And this has driven all manner of recombinant innovations. The Internet has been a positive in many profound ways.

But high diversity in the overall network has the potential to create paradoxical effects in local networks. This is as true in the digital world as in the social world. At a cosmopolitan university such as the University of Kansas, it can lead to homogenous friendship groups. At a mixer purposely set up to encourage people to mingle, it led to fine-grain assorting.

These insights help us grasp one of the defining paradoxes of the modern age: echo chambers. For all its promise of diversity and interconnection, the Internet has become characterized by a new species of highly homogenous in-groups, linked not by the logic of kin or nomadic tribe, but by ideological fine-sorting. This is a thoroughly digital incarnation of the segmentary dynamics of the Neolithic era, with information circulating within groups rather than between them. In many cases, echo chambers are nothing to worry about. If you are interested in fashion, you want to join a forum where you can converse with like-minded others. It would undermine your enjoyment if people kept posting about architecture, or soccer, or fitness. Diversity is not just redundant but irritating in such forums.

But when one is seeking to become informed on complex subjects such as politics, echo chambers are inherently distorting. By getting their news from Facebook and other platforms, where friends share cultural and political leanings, people are more exposed to those who agree with them, and evidence that supports their views. They are less exposed to opposing perspectives. The dynamics of fine-sorting can be magnified by a subtler phenomenon: the so-called filter bubble. This is where various algorithms, such as those inside Google, invisibly personalize our searches, making it more likely that we'll see more of what we want to see, further limiting our access to diverse viewpoints. This is the digital equivalent of the Bahns experiment, but at a higher level of gearing. The sheer interconnectivity of the Internet has facilitated enhanced political fine-tuning.

The precise extent of echo chambers is a matter of some debate, with different studies pointing in slightly different directions. The mathematician Emma Pierson analyzed how the troubles of Ferguson, Missouri, were covered on social media in 2014, after police officer Darren Wilson shot and killed a young Black man, Michael Brown. She found two distinct clusters. "Blue tweets" expressed horror at Brown's death and criticized the oppressive police response, while the "red tweets" argued that the policeman was being scapegoated and the protesters were looters. As Pierson puts it:

The red group says they would feel safer meeting Darren Wilson than Michael Brown, and says that Brown was armed when he was shot; the blue group sarcastically contrasts Darren Wilson with the unarmed Michael Brown. The red group talks about mob justice and race baiting; the blue group talks about breaking the system. The red group blames Obama for exacerbating tensions and forcing the Missouri governor into declaring a state of emergency; the blue group says the state of emergency must not be used to violate human rights.

Perhaps most tellingly of all, these two groups had virtually no interaction with each other. They were only seeing tweets from people who agreed with them, a demonstration of how the segmentary dynamics of the Internet can filter information. "When it comes to Ferguson, two groups with very different political and racial backgrounds ignore each other," Pierson writes. "This seems likely to cause problems, and in fact it does. For one thing, the two groups think drastically different things."[12]

Other studies led by Seth Flaxman of Oxford and the Pew Research Center offer a different lens on the digital world. When you look at overall Internet use, digital users have higher average exposure to the views of their own side, but nevertheless get to see the views of opponents, too. Perhaps that is not surprising. Even in the clan systems that emerged after the agricultural revolution, the various in-groups were not completely shut off from each other.

But what is fascinating—and broadly acknowledged by almost all scholars—is what happens when exposure does take place. Now, you might have thought that by hearing the views of opponents and seeing the evidence from the other side, opinions would become less extreme. Views would become more nuanced. In fact, the opposite happens. People become *more* polarized. In Pierson's study, for example, the limited interaction between red and blue tweeters was explosive. She writes:

> When the red and blue groups did talk, it often wasn't pretty.
> Consider the things said by members of the red group to one
> of the most influential members of the blue group—DeRay
> Mckesson, a school administrator who has played a central role in
> organizing protests. They described him as a "commie boy" who
> spread hate . . . saw "value in racist drivel," was armed with "guns
> and Molotov cocktails," and should get his "meds adjusted."

---

[12] A 2019 study led by Ana Lucía Schmidt, a computational social scientist based in Italy, came to broadly similar conclusions. She analyzed 376 million Facebook users' interactions with 900 news outlets and concluded that "selective exposure drives news consumption. . . . We find a distinct community structure and strong user polarization."

A different study concluded that "segregation of users in echo chambers might be an emerging effect of users' activity on social media."

A study led by Christopher Bail of Duke University found a similar pattern. He recruited 800 Twitter users to follow a bot that retweeted the views of high-profile people from across the political spectrum. What happened? Far from becoming more balanced, the Twitter users became more polarized. This was particularly true for Republicans, who became more conservative. It was as if exposure to different views confirmed their prior convictions.

To understand what is going on, and to fully glimpse the internal logic of echo chambers, we need to draw a subtle distinction between echo chambers and information bubbles. As the philosopher C. Thi Nguyen notes, information bubbles are the most extreme form of isolation, where people on the inside see only their side of the argument and *nothing else*. These kinds of social groups have rarely existed in modern history except in cults and other "walled institutions." Echo chambers, Nguyen argues, are different. They may cut some people off from alternative views through informational filtering (research by digital scholars Elizabeth Dubois and Grant Blank found that 8 percent of people in the United Kingdom have such biased media exposure that they experience a distorted version of reality), but their distinctive feature is that they have not one filter, but two.

What is the second filter? We will call it *epistemic walls.*

III

In their scholarly book *Echo Chamber,* Kathleen Hall Jamieson and Joseph N. Cappella, two experts on politics and the media, examine the core logic of political polarization. They do so through the prism of Rush Limbaugh, a hugely successful conservative commentator whose radio show has a cumulative weekly audience of around 13.25 million unique listeners.

Jamieson and Cappella note that Limbaugh doesn't seek to persuade his audience to cut themselves off from alternative voices. This would be impossible in such an interconnected world. Instead, he seeks to *delegitimize* alternative voices. He attacks the integrity of those who offer different views, and defames their motives. His insistence is not (just) that opponents are wrong, but that they are malicious. He argues that the mainstream media expresses a liberal bias and has set out to destroy Limbaugh and his followers because they can't abide the truth he speaks. Jamieson and Cappella write, "The conservative opinion hosts underscore the notion that the mainstream media use a double standard that systematically disadvantages conservatives and their beliefs. They argue that Limbaugh seeks to discredit all other sources of information, along with political opponents, through the techniques of 'extreme hypotheticals, ridicule, challenges to character, and association with strong negative emotion.'"

Now we can begin to glimpse the subtly different properties of information bubbles and echo chambers. With the former, informational borders are hermetically sealed. People on the inside only hear people who are co-inhabitants of the bubble. This creates distortions, but it also confers fragility. The moment a member of the in-group is confronted with outsider opinions, they are likely to question their beliefs. The way to burst an information bubble, then, is through exposure. This is why cults take such lengths to deny insiders exposure to different voices.

Echo chambers, with their additional filter, have fundamentally different properties. People on the inside hear more opinions from the in-group, but these views tend to become *stronger* when exposed to opposing opinions. Why? Because the more opponents attack Limbaugh, the more they point to the errors in his opinions, the more it confirms the conspiracy against him. Opponents are offering not new insights, but fake news. Each piece of evidence against Limbaugh is a new brick in the wall separating the in-group from outsiders. As Nguyen puts it:

What's happening is a kind of intellectual judo, in which the power and enthusiasm of contrary voices are turned against those contrary voices through a carefully rigged internal structure of belief. Limbaugh's followers read—but do not accept—mainstream and liberal news sources. They are isolated, not by selective exposure, but by changes in who they accept as authorities, experts, and trusted sources. They hear, but dismiss, outside voices.

Perhaps the clinching point is that trust is an essential ingredient of belief formation. Why? Because we don't have the time to check the evidence for everything, so we have to take some things at face value. We trust doctors, pharmacists, and teachers. Even experts trust other experts, taking their data and outputs as inputs for their own deliberations, because checking from first principles is virtually impossible. The world of information, somewhat like commerce, is presupposed by trust. As Nguyen puts it:

> Ask yourself: Could you tell a good statistician from an incompetent one? A good biologist from a bad one? A good nuclear engineer, or radiologist, or macro-economist, from a bad one? . . . Nobody can really assess such a long chain for herself. Instead, we depend on a vastly complicated social structure of trust. We must trust each other, but, as the philosopher Annette Baier says, that trust makes us vulnerable.

It is this epistemic vulnerability that echo chambers exploit. By systematically undermining trust in alternative views, by defaming those who offer different insights and perspectives, they introduce a filter that distorts the belief-formation process itself. Alternative views are dismissed not after consideration, but on contact. Facts are rejected even as they are offered. Perspectives and evidence are repelled like a magnet repelling iron filings. Nguyen writes, "Echo chambers operate as a kind of social parasite on our vulnerability. . . . An information bubble is when you don't hear people from the other side. An echo chamber is what happens when you don't trust people from the other side."

This is not about conservative radio hosts, still less conservatism itself. There are echo chambers not just on the Right, but on the Left and, indeed, beyond politics. "The world of anti-vaccination is clearly an echo chamber, and it is one that crosses political lines. I've also encountered echo chambers on topics as broad as diet (Paleo!), exercise technique (CrossFit!), breastfeeding, some academic intellectual traditions, and many, many more."

It is the twin filters of information and trust that create an unusually resilient form of in-group cohesion. Where an information bubble is inherently fragile, echo chambers on both sides of the political spectrum are reinforced by the mutual exposure of alternative views, driving polarization and leading to competing (and often contradictory) claims of fake news. Each side thinks the other is living in a post-truth age. As Nguyen says, "Here's a basic check: Does a community's belief system actively undermine the trustworthiness of any outsiders who don't subscribe to its central dogmas? Then it's probably an echo chamber."

IV

Derek Black grew up in an echo chamber both on- and offline. At six, he was sent out for Halloween as a white Power Ranger. A little later, his father hung a poster of the Confederate flag on his wall. He started attending white supremacist conferences at the same time, hearing adults talking about the inherent intellectual inferiority of Black people. Saslow writes that "Derek was socialized on Stormfront, and he began spending his nights in the private chat room as soon as he could type. After Derek finished third grade, Don and Chloe pulled him out of school, believing the public system in West Palm Beach was overwhelmed with an influx of Haitians and Hispanics."

After that, he was homeschooled, imbibing yet more supremacist ideology, with consistent exposure to the politics of racism. The Black family lived in West Palm Beach, but their house was like an island, surrounded by vegetation that Don allowed to grow high and wild. No visitors were allowed into the house except fellow white supremacists and family members. It would be easy to assume, then, that Derek's extremist views were sustained through social isolation. He didn't question his own beliefs because he was not exposed to any others. In fact, although Derek did lead an unusual life, he was not in a cult. David Duke, his godfather and the de facto leader of American white nationalism, did not seek to prevent him from hearing contrary opinions. Neither did his parents. This was not, to use our previous terminology, an information bubble.

No, this was an echo chamber. Duke and Black did not bar alternative sources of information; rather, they systematically undermined his trust in them. Nonsupremacists, of all kinds, were positioned as deceitful, members of a liberal establishment intent on selling out white Europeans to immigrants and Jews, people who couldn't tolerate the expression, still less the adoption, of the "reasonable demands" of the far right.

This explains why Derek's views, far from softening as he was exposed to the Internet, various TV stations, and other sources of information, actually hardened. To him, these contrary voices were not expressing reasonable opinions, but were peddlers of fake news. They were duplicitous manifestations of a politically correct establishment. Saslow puts it this way: "He was impervious to feedback from strangers. His critics were nothing more to him than an anonymous chorus on the other side of a curtain—a circus of 'usurpers' and 'Neanderthals.' . . . If he didn't respect them, why would he care about their opinions?"

At the age of eighteen, Black left home to go to university. He chose the New College of Florida, one of the top colleges in the state. His chosen subjects were German and medieval history, "which he had always associated with the glorious dominance of white Europeans. His parents reminded him that, ultimately, they hoped he would make history and not just study it." His father was not in the least bit worried that his extremist views might be moderated by contrary voices. When a caller to the radio show asked about Derek finding himself within "a hotbed of multiculturalism," Don laughed. "It's not like any of these little commies are going to impact his thinking. If anyone is going to be influenced here, it will be them."

But the New College of Florida is unusual. It is *small*. In total, there are only eight hundred or so students. At a large university, Derek might have found a critical mass of friends on the far right of politics. He might have constructed an ideologically similar network. In a small college, there was no such scope for political fine-tuning. Paradoxically, he was about to find himself more exposed to contrary opinions than ever before. The information filters were about to disappear *completely*.

On his very first day, he bumped into Juan Ellis, an immigrant from Peru with a wispy beard and long sideburns. Derek had hardly spent any time with anyone Hispanic before. They chatted at length about life and more. A few days later, he started playing guitar in the courtyard, and noticed a student wearing a yarmulke sitting down to listen. Matthew Stevenson was the only Orthodox Jew on campus and he started to sing along, smiling.

Derek made the decision early on to conceal his political identity. He was careful never to talk about politics, or at least never to hint at his true beliefs. He didn't want to become socially isolated at college. He would spend time with his fellow students in the evening, chatting about history, or languages, or music, then leave his dorm early in the morning to call in to his radio show, broadcasting far-right sentiments on the airwaves. Nobody twigged what was going on. Saslow writes, "On the air, he repeatedly theorized about the 'criminal nature of blacks' and the 'inferior natural intelligence of blacks and Hispanics.' He said President Obama 'was anti-white culture,' a 'radical black activist,' and 'inherently un-American.'"

After a year, Derek's views had not budged. The trust filters were taking the strain of sustaining his extremist ideology. He remained the great hope of the far right. Besides, the perennial slur that he had no mind of his own had always bugged him. He was pleased that his convictions had remained robust in such an alien environment. "Derek hated the suggestion that he'd simply been indoctrinated with his family's racial convictions; no idea was more insulting to him." After his first semester, he flew to Germany for a four-month sabbatical in an immersive language school. He visited his godfather, David Duke, and continued to read up on racial ideologies. He was three weeks into his visit when he logged in to the student blog back at New College to chat with friends and to catch up on the news.

A few days previously, at 1:56 a.m., a student studying far-right extremism had come across a photo of a youngster with long red hair and a cowboy hat on a far-right website. He was stunned. "Have you seen this man?" he posted, with a picture beneath it. "Derek Black. White supremacist, radio host . . . New College student?"

Within hours, it had become the most blogged post in the history of the college.

Black knew what was coming. When he returned to college, he was ostracized by former friends. "I just want this guy to die a painful death along with his entire family; is that so much to ask?" one wrote on a student message board known as "the forum." Another wrote, "Violence against white supremacists will send a message that white supremacists will get beat up. That's *very* productive." He was confronted by fellow students leaving a party, and had to be dragged away by someone fearing he was about to get punched. People vandalized his car. Others shouted expletives. At one point, students shut down the school for a day in protest at Derek's presence.

To Derek, this merely confirmed what he had learned from Duke and his father. The liberal establishment was out to get the far right. They couldn't bear alternative opinions. They wouldn't even allow them to be expressed. They were the bigots and censors. It was white nationalists who had the scientific and moral arguments on their side. Black showed his defiance by organizing an international conference for members of Stormfront with the theme "Verbal Tactics for Anyone White and Normal." "Come and learn how to stand strong against the enemy's abuse," he announced on radio. He booked a dozen keynote speakers—including his father and Duke, two of the most eloquent advocates of white nationalism.

Derek "obsessed over each detail, from the colors of the conference logo to the sandwiches in the pack lunches." Even before he started his speech to open the conference, the assembled extremists from Europe, Australia, and Canada stood to applaud. He found himself in the bosom of the community.

"Gen-o-cide," was David Duke's opening remark. "Say it with me now. This is the murder of our very genes. Repeat that over and over." Don, Derek's father, spoke last, to yet more resounding applause, Derek coming to stand alongside him on the stage.

"Staying on the genocide message demoralizes/embarrasses anti-whites," Derek wrote on the Stormfront message board. "Stay on the offensive, because you are right." His destiny as the future force for white nationalism seemed surer than ever. Then, a few days later, everything changed.

V

Matthew Stevenson has black hair, a short beard, and bright eyes. His manner is calm, his face friendly. He grew up in Miami, Florida, in a Jewish family, started attending the Kabbalah Centre at the age of eight, and grew in the Jewish faith. At fourteen, he started wearing the yarmulke. His upbringing was, at times, tough, not least because of his mother's alcoholism. She attended a treatment facility when he was young, and he accompanied her to AA meetings from the age of seven. "It was a huge education. You meet all sorts of people: rich, poor, white, Black. You hear incredible stories of how people have reached rock bottom, but managed to find a way back. You develop empathy."

Matthew is the Orthodox Jew whom Derek encountered on his first day at New College, wearing a yarmulke and singing along to Derek's guitar in the courtyard. I interviewed Matthew on a sunny winter afternoon and found him to be quizzical and thoughtful, a young man who had learned a great deal at those AA meetings, not least people's capacity for change.

"It was quite difficult to process that information," he told me about the day he found out about Derek's white supremacist beliefs. "When we first met, I had no idea about his ideology. We just enjoyed chatting together and hanging out. We weren't best friends, but we knew each other and liked one another's company. When the news broke, I was as shocked as everyone else."

Matthew was already aware of the Stormfront website. Like other students worried about the rise of the far right, he had surfed the site to try to make sense of the sentiments driving the rising tide of hate crimes. "When I heard the news about Derek I went back to the site, and looked to find his postings. It was pretty sobering." One of the posts from Derek that Matthew uncovered said, "Jews are NOT white. They worm their way into power over society. They manipulate. They abuse."

Many of Matthew's fellow students rejected Derek instantly, while others verbally abused him. The student forum was ablaze with shock and scandal for many weeks and months. But Matthew pondered Derek's upbringing, the virulent culture of white nationalism in which he had been socialized, and reflected on just how easily any young person might end up with racist views in that milieu. Matthew says:

I knew that it was very unlikely he'd spent a lot of time around other people. He didn't have a lot of Black family friends or Jewish family friends. I couldn't put my hand on my heart and say that I wouldn't have become a white nationalist in that subculture. I felt that the right thing to do was to reach out to him. I knew from AA how people can change, often dramatically.

Every Friday, Matthew hosted a Shabbat dinner for friends. It had started out as a small group but had grown to include Christians and atheists, and was something of a fixture in the social life of the tiny campus. Often, up to fifteen people would pack into his dorm room to eat honey-and-mustard-glazed salmon and challah bread. It was a wonderful way to build friendships and share ideas.

Just a few days after the Stormfront conference, with Derek now back on campus, Matthew sat down to write a text message to the campus's white supremacist. "Hey," the message said. "What are you doing on Friday night?" Matthew followed up on the Friday afternoon: "Looking forward to seeing you tonight." Derek, more isolated than ever before, accepted the invitation. "I wasn't getting many social invitations at the time!" he later said. Matthew recalls:

At first, it was a little awkward. Neither of us knew how it was going to go. I had asked the other two guests [most of the usual attendees had refused to turn up because of Derek's presence] not to bring up politics. After a few minutes, it flowed smoothly. . . . He is an intelligent guy. He came the next week, and the next. Frankly, I enjoyed his friendship.

Matthew avoided politics. He realized that such a hot topic could lead to sharp exchanges, not least with the other students, many of whom had started to return to the weekly dinner after boycotting Derek's first appearance. Matthew doubted that such an exchange would change anyone's views—at least, not at first. He knew that before a meaningful dialogue could take place, he needed to establish something else: trust.

They chatted about early Christianity, languages, monasticism. Derek was consistently impressed with the scope of Matthew's knowledge. Matthew, for his part, thought that Derek was one of the smartest guys he had met. These were two of the top students in the college, and they were building an ever-stronger connection. They laughed together. They learned together. More attendees started to return to the Shabbat dinner and began forming bonds with Derek, too. Brick by brick, the epistemic wall was being dismantled.

When one evening Alisson Gornik, another attendee at the dinners, brought up his political views when they were chatting on their own, Derek listened. They discussed the foundations of white nationalism: the idea that Blacks are, on average, less intelligent than whites, that they are more predisposed to crime, that there are immutable biological differences between races. Derek believed in the robustness of these pseudoscientific foundations. He was sincere when he said that he thought that minorities should be repatriated. He believed that it was better for Blacks and whites.

Alisson came back with scientific papers that challenged the statistical basis of these racist claims. Derek had heard of such papers before but had never engaged with them. Why bother with the duplicitous data introduced by an untrustworthy liberal and scientific establishment? Why spend time on information that has been rigged to obtain prejudged results?

Now, he found himself reading them with a more open mind. He saw papers that showed that IQ differences could be explained by cultural biases. He read about stereotype threat, how first-generation immigrant children performed better at school, on average, than American students, about the basis of human genetic variation and its implications.

He had long sincerely argued that whites were discriminated against in modern America but now found himself face to face with data on the lack of Black representation in state government, how white employees are more likely to be promoted than equally qualified Black employees, how Black students are twice as likely to be suspended from school for the same offence, how Black people are twice as likely as white people to work for the minimum wage in a given job, how Black applicants are significantly less likely to be invited for interview even with identical qualifications.

Was this really a country rigged in favor of minorities and against white people?

Derek's life, his childhood memories, his sense of identity, was bound up with white nationalism. So were his family, his friends, his in-group. But the foundations of his beliefs were being dismantled, not because he had never been exposed to contrary evidence, but because he hadn't engaged with it. Slowly but surely, he came to the realization that the evidence didn't support the ideology of white nationalism, even if he knew that admitting so publicly would cause ructions across the movement and jeopardize his relationships, most significantly with his mom and dad. One evening, he sat down and started to type:

A large section of the community I grew up in believes strongly in white nationalism, and members of my family whom I respect greatly, particularly my father, have long been resolute advocates for that cause. From a young age I observed my dad sacrifice dearly for his commitment—a conviction stemming from nowhere else than ardent resolve in the rightness of the cause. I was not prepared to risk driving any wedge in those relationships and I did not believe that was necessary.

The number of changes in my beliefs during the past few years, however, has amounted to a shift that I think needs to be addressed. It is impossible to argue rationally that in our society, with its overwhelming disparity between white power and that of everyone else, racial equity programs . . . represent oppression of whites. . . . Particularly bizarre to me is the determination of Jewish social domination. . . . It is an advocacy that I cannot support, having grown past my bubble, talked to the people I affected, read more widely, and realized the necessary impact my actions had on people I never wanted to harm.

He then found the email address of the Southern Poverty Law Center (SPLC), the civil rights group that had scrutinized the activities of his father for so many decades, and pressed send.

VI

The defining error in the contemporary analysis of the post-truth age has been the conflation of information bubbles and echo chambers. The former concept seeks to explain people's extremist beliefs via distorted exposure. The idea is that when people are denied access to diverse views and evidence, they are more likely to cleave to extremist beliefs and ideologies. As the legal scholar Cass Sunstein argued in a highly influential essay:

> Although millions of people are using the Internet to expand their horizons, many people are doing the opposite, creating a Daily Me that is specifically tailored to their own interests and prejudices. . . . It is important to realize that a well-functioning democracy—a republic—depends not just on freedom from censorship, but also on . . . unsought, unanticipated, and even unwanted exposures to diverse topics, people, and ideas. A system of "gated communities" is as unhealthy for cyberspace as it is for the real world.

Although this analysis sounded plausible, it struggled to survive empirical scrutiny. When the evidence showed that many at the extreme ends of politics are, in fact, exposed to contrary opinions, but seemed impervious to them, a new set of explanations arose. These focused on psychology (people are just too lazy to engage with contrary opinions) or accusations of downright irrationality. The idea seemed to be that many people had lost faith in truth itself.

Grasping the properties of echo chambers offers a far more plausible explanation. The problem isn't that people have become trapped in cult-like information bubbles, nor has there been an epidemic of irrationality. No, the problem is subtler. When outside sources of information have been systematically discredited, the belief-formation process itself undergoes distortion. In a world where trust is, in a certain sense, prior to evidence, this can be perilous. As Nguyen puts it:

> Echo chambers are structures of strategic discrediting, rather than bad informational connectivity. Echo chambers can exist

even when information flows well. In fact, echo chambers should hope that their members are exposed to media from the outside; if the right disagreement reinforcement mechanisms are in place, that exposure will only reinforce the echo chambers' members' allegiance. We ought not conclude then, from data that epistemic bubbles do not exist, that echo chambers also do not exist.

DEREK BLACK AND MATTHEW STEVENSON ARE, today, two of the most eloquent voices arguing against political polarization. Neither has a problem with divergent opinions, even powerfully argued ones, but they do worry about the character attacks, allegations of fake news, and a broader breakdown of trust in political opponents. They have shared a stage on popular TV programs, at youth events, even at corporations seeking to understand the post-truth age. Matthew—who is studying for a PhD in economics and mathematics—now helps a charity seeking to promote community understanding, while Derek is completing his PhD in history. Derek's Twitter handle? "Unexpected advocate for antiracism."

Derek's life wasn't easy after he emailed his statement to the SPLC retracting his political views. The storm in white nationalism was intense. Don, his father, initially thought the email had been sent by an impostor, and his mother didn't want to speak to him. As for Duke, he conjectured that Derek was suffering from Stockholm syndrome—he had been effectively taken hostage by the liberal elite and was experiencing empathy for his captors. Derek has said:

Immediately, my dad called me and said he thought my email had been hacked because I had not shared with them this process. And so my condemning [white nationalism] came as a real shock to him. I am honestly not super proud of the way I did it. I feel like I should have warned him and talked to him a bit more. And we had days of very intense conversations where it was not clear whether we would continue to speak at all.

Matthew tells me:

I think the first few months were the most difficult for Derek. His social life and identity were bound up in that ideology. It has taken a lot of readjustment. But Derek's transformation confirmed what I learned when I went with my mother to Alcoholics Anonymous. People are capable of change, if you gain their trust. People start to listen to what you are saying when there is a real relationship, rather than just rejecting what you say out of hand.

That is the problem today: the mudslinging that characterizes competing political groups. This is true of many, not just the hard right. . . . It makes dialogue almost impossible.

Philosophers have a particular term for serial assaults on personal integrity. The *ad hominem* argument is defined by one reference source as "a fallacious argumentative strategy whereby genuine discussion of the topic at hand is avoided by instead attacking the character, motive, or other attribute of the person making the argument, or persons associated with the argument, rather than attacking the substance of the argument itself." A paper by the Finnish philosopher Jaakko Hintikka finds that the fallacy was first discussed by Aristotle in his book *On Sophistical Refutations*. It has been a staple of philosophical treatment ever since, particularly in the work of John Locke. "It may be worth our while a little to reflect on . . . arguments that men, in their reasoning with others, do ordinarily make use of to prevail on their assent, or at least to awe them into silence," he writes.

Psychological research reveals the power of the ad hominem. One recent paper published in the Public Library of Science surveyed 39 college students and 199 adults. They found that when you attack someone's character it undermines faith in their conclusions as strongly as when you identify actual evidence questioning the basis of those claims. Playing the person rather than the ball *works*. In this sense, the ad hominem argument represents what economists call a *free rider problem*. All citizens benefit from the trust that is central to the functioning of democratic institutions, but the importance of trust offers politicians an incentive to impugn the integrity of opponents, thus benefiting in electoral terms, but weakening the epistemic fabric on which the collective intelligence of any democracy depends. Trust itself, as a consequence, starts to become fragmented.

The ad hominem argument is not always fallacious, of course. If someone has consistently lied, or has a conflict of interest, then drawing attention to that is legitimate. The problem is when a person's character is attacked not because of wrongdoing, but because she is an opponent—when a viewpoint is taken as prima facie evidence of bad faith. This kind of epistemological tribalism is not an impoverished form of informed debate, it is the antithesis of informed debate.

Many Ancient Greek philosophers, not least Socrates himself, argued that the good functioning of a democracy is inextricably linked to the quality of its deliberations. It is only by testing ideas, by examining evidence, that we reach reasoned decisions. This is the lesson that emerges from our analysis of diversity science, too, as well as from formal theories that probe the conditions under which democracy leads to wise outcomes.[15] And this is precisely why Socrates believed that it was crucial for citizens to be able to detect and punish false reasoning. For Stevenson, one of America's most fascinating modern citizens, this offers a ray of hope. "If public figures knew that constantly impugning the character of oppo-

---

[15] In her excellent book *Democratic Reason: Politics, Collective Intelligence, and the Rule of the Many* (Princeton University Press, 2012), the Yale political scientist Hélène Landemore pro-

nents would lose them credibility with their own side, they might engage on the evidence instead," he told me. "That would improve the tone of debate, and the quality of deliberations. If someone automatically attacks the trustworthiness of opponents, he should forfeit trust in himself."

As for Derek Black, after the initial standoff with his parents, there has been a partial reconciliation. "We text, we call every now and again, I have been back three or four times for a day or two in the last five or six years," he has said. "Being able to communicate was more important than our political differences.... My parents should take some credit for that." It is difficult to predict where their relationship will end up, but it is not impossible that the guitar-playing student whose epistemic walls were dismantled by Stevenson will, in turn, help dismantle the walls of his father. After all, there is not merely trust between Derek and Don, but also love. Indeed, their discussion might one day lead to the most dramatic conversion of all: that of one of the highest-profile white supremacists in modern political history.

"They say that distrust is contagious," Matthew says. "Sometimes trust can be contagious, too."

---

vides a powerful defense of democracy from the perspective of collective intelligence. Under many conditions, she shows that many minds will reach better decisions than oligarchies, dictatorships, and military juntas. Classic examples include Condorcet's jury theorem, developed by the Marquis de Condorcet and published in his 1785 work *Essay on the Application of Analysis to the Probability of Majority Decisions*.

BEYOND
AVERAGE

I

So far in the book, we have examined the dangers of homophily, dominance dynamics, and echo chambers. We have looked at the power of an
outsider mindset and recombinant innovation. We have also seen how an
understanding of diversity shines a light on everything from the failures
of the CIA to the benefits of constructive dissent.

In this chapter, we are going to look at diversity from a fresh angle.
There is a conceptual flaw in the way we think about ourselves as human
beings. This flaw has infiltrated many branches of science, and it can

undermine the influence of diversity, preventing institutions and societies from reaching their full potential.

To define this fallacy, and why it matters, will take a bit of digging. We will start with one of its most confounding manifestations, one that we're all familiar with: the changing and often contradictory advice on diet and nutrition. This may seem to have little to do with the science of diversity, but as we will see, it shines a light on an important facet of our world. One with vast implications for rebel ideas and for our future.

ERAN SEGAL WAS CONFUSED. In fact, it is probably more accurate to say that he was befuddled. For a brilliant scientist who had earned his PhD from Stanford University, this was not an agreeable feeling. The source of his confusion will be familiar to anyone who has spent more than a few moments contemplating diet and nutrition. The question of what to eat is vital to health and longevity, yet the evidence remains confusing.

As an undergraduate, Segal played lots of handball and ate healthily, but was forty to fifty pounds overweight. When he got together with Keren, his wife, whom he met at a party at twenty-three, his confusion heightened. Keren would soon become a clinical dietician, armed with the latest science. She cooked healthy meals, with plenty of fresh vegetables, largely following the guidelines of the Academy of Nutrition and Dietetics, but Eran wasn't shedding any pounds.

"I decided to stop following the guidelines and look at the scientific evidence on which they were based," Eran says when I chat with him and Keren. "It wasn't what I expected to find. A lot of studies were based on small samples of people. And a lot had been funded by food corporations, which made one skeptical about the results. It wasn't as rigorous as I expected."

As Eran is talking, Keren's face breaks into a smile. "He is an easygoing person most of the time, but he cares a lot about data. This really got to him."

Perhaps Eran's biggest surprise was that the evidence seemed contradictory. Some studies called for low-fat diets. Others for high-fat diets. Popular books eulogized a Paleo diet, or a Mediterranean diet, or an Asian diet, or some combination of the three, or some other craze that had gripped consumers and then faded away, perhaps to return in another form, with a few twists.

Or take carbohydrates. Some evidence suggests that low-carb, high-fat diets can improve health, while others suggest that the best diets are low-fat, high-carb. There is evidence for both sides, which, in a sense, is evidence for neither. This isn't just confusing; it is downright mysterious. It caused Segal's frustration to harden into a deeper kind of fascination.

Segal has a quizzical face, and his eyes flash as he talks about his intellectual journey. In many respects, he is like anyone else who has been befuddled by dietary advice. He lives with Keren, three kids, a dog called Snow, and a cat called Blue, and he does his best to live responsibly. But he had one sizable advantage in getting to the bottom of the conundrum: a world-class background in computation (in his twenties, he was the winner of the prestigious Overton Prize, awarded annually for outstanding accomplishment by a young scientist) and a post at the Weizmann Institute of Science, one of the top academic institutions in the world. Segal says:

> It is no surprise that people are confused. In 2012, the American
> Heart Association and the American Diabetes Association
> suggested that people should drink diet soda for weight loss and
> health. There was then a huge rise in diet soda consumption,

even though further research showed the opposite. In 1977, the U.S. government said that fat was bad and fiber good, so people reduced fat and increased fiber. At much the same time, obesity tripled in men and doubled in women.

This last point shows that Segal's concerns are of more than theoretical interest. Diet has become a huge public issue. If you live in the United States, you have a nearly 70 percent chance of being overweight and 40 percent chance of being obese. In the United Kingdom, the stats are similar. Obesity rates around the world have more than doubled since 1980, and in 2014 more than 1.9 billion adults, 39 percent of the world's population, were overweight, with 600 million obese. Segal says, "The obesity epidemic can't have been helped by the fact that so many people are baffled by the advice. People who go on diets often put their weight back on. In fact, there is a lot of evidence to suggest that dieting is associated with piling pounds on rather than taking them off."

A study conducted on the dieting reality show *The Biggest Loser* showed that contestants lost huge amounts of weight through exercise and calorie counting. And yet this drastic loss caused their metabolic rates to plummet to such an extent that, six years later, their metabolisms were so slow that they were not able to eat the same calories as people of the same weight who had never been obese. Scientists call this *persistent metabolic adaptation.* And yet this was just one of dozens of anomalies that Eran found as he probed the science. "There were certain agreed facts, such as that any diet should include fat, salt, protein, fiber, vitamins, and minerals," he says, "but beyond that, almost anything seemed to go."

Another frustration emerged when he got into marathon running in his thirties and looked to see if diet could help him improve his times. Sure enough, the dietary advice offered to runners is every bit as contradictory as that given to everyone else. He says:

When I first started running, the big thing was "carb loading" the night before a big race. It was an axiom of marathon running. I would typically eat three bowls of pasta the evening before a race, and then a few dates or energy bars thirty minutes before a run. At first, I didn't question this advice, but after a while, I decided to take a closer look.

The more he looked, the more perplexed he became. Some studies treated carbs as all the same; other suggested that there were "good" carbs and "bad" carbs. One study claimed that eating dates thirty to sixty minutes before running energized some runners, while others felt so exhausted that they had to stop a few minutes into their runs.

"I decided to conduct an experiment on myself," Eran says. "One night, instead of eating multiple bowls of pasta, I ate a salad with lots of fat sources like avocado and nuts and tahini. The following morning, I did a twenty-mile run without eating anything at all." This contradicted mainstream advice. Indeed, many dieticians would have described it as self-sabotage. Yet he felt stronger, fitter, and more primed than ever. He went on: "My energy levels were higher than with carbs, and my postrun hunger pangs disappeared. I inferred that my body had switched from burning carbs to burning fat, altering my energy and hunger."

In his thirties, he achieved his ambition of running a marathon in under three hours in Paris. In 2017, he ducked under three hours again, this time in Vienna. And yet he still hadn't cracked his ultimate ambition of making sense of dietary science. "I couldn't let it go," he says. "This was a mystery that had to be solved."

II

In the late 1940s and early 1950s, the U.S. Air Force confronted a mystery of its own. It was the early era of jet-powered flight, when engineering was supposed to have reached unprecedented levels of reliability, but the Air Force was enduring incident after incident. There were crash landings, unplanned dives, and much else besides. "The problems were so frequent and involved so many different aircraft that the Air Force had an alarming life-or-death mystery on its hands," writes Todd Rose, a Harvard academic. "It was a difficult time flying," one retired airman said. "You never knew if you were going to end up in the dirt."

To get a sense of the scale of the problem, consider U.S. Air Force official records for February 1950. On the first day of the month, safety incidents were reported for Charles L. Ferguson in a C-82 Packet, a twin-engined, twin-boom cargo aircraft, Medford Travers in a P-51 Mustang, a long-range, single-seat fighter, Malcolm W. Hannah in a T-6 Texan, a trainer aircraft, and Herman L. Smith in a Boeing B-29. Others who suffered incidents included Harry L. McGraw, William K. Hook, and George T. Shuster. Just to be clear, these all happened *on the same day.* But February 1, 1950, wasn't an outlier, it was a typical twenty-four hours. On the second day of that same month, there were four safety incidents, on the third there were seven, and on the fourth there were four more. On the fourteenth of that month, there were *sixteen separate incidents.* In total, 172 incidents were recorded that month.

What was going on? The problem didn't seem to be with the mechanical or electronic systems of the planes. These had been thoroughly tested by engineers and found to be in good working order. Yet the incidents didn't appear to be caused by a sudden deterioration in the skill of pilots, either. These were well-trained professionals, highly regarded within the industry.

But if it wasn't engineering or skill, what else could it be? Into the heart of this mystery stepped a Harvard graduate who specialized in physical anthropology. Lieutenant Gilbert S. Daniels was not a conventional airman: he was quiet and softly spoken, methodical and scientific, and listed gardening among his hobbies. In persona and interests he was not dissimilar to Eran Segal. And Daniels had a strong hunch. He believed that the problem wasn't with engineering or pilot judgment, but with the *design of the cockpit itself.*

The background is instructive. The cockpit design had been standardized in 1926 by the U.S. Air Force after tabulating the dimensions of hundreds of airmen. This process uncovered the average physical traits of pilots, which were then used to determine the height of the chair, the distance to the pedals and stick, the height of the windscreen, the shape of the helmets, and so on. Some within the Air Force initially pondered if pilots had become bigger since 1926, making it difficult to operate the controls. Might that be the explanation for all the crashes? Daniels had a different hunch. He reckoned the problem wasn't that the average airman had grown, but that the notion of the average airman was itself defective. Perhaps there was no such thing as an average airman.

In 1952, Daniels had the opportunity to test his intuition. He led a project at Wright-Patterson Air Force Base, which set out to measure the physical dimensions of pilots. Daniels threw himself into the task, carefully tabulating 4,063 pilots on 140 dimensions of size. These included "thumb length, crotch height, and the distance from a pilot's eye to his ear." He then calculated the average for the ten dimensions that he deemed most important when it came to cockpit design. In other words, he was tabulating the dimensions of the average pilot.

But how many airmen would conform to this average? Daniels was quite generous in the way he calculated the latitude. If the measurement of a given pilot was within the middle 30 percent of the range of values for a given dimension, he was deemed to be average. For example, the average height of the airmen was five feet nine. So Daniels considered any particular pilot to be average in this dimension if he fell within the range of five feet seven to five feet eleven.

Now, most experts in the military assumed that the majority of pilots would be in the average range across most of these ten dimensions. Indeed, this conclusion seemed obvious. After all, the average was calculated from that very sample of airmen! Moreover (as Rose points out), these airmen had been preselected precisely because they conformed to the basic specification demanded by the Air Force. They would never have hired a five-foot-four pilot in the first place.

But what happened? How many pilots were within the average range across the ten dimensions? Zero. *Not a single airman*. From a cohort of more than four thousand, none were average. Daniels's hunch was emphatically confirmed. The problem wasn't that the average pilot had got bigger since 1926. The problem was that there was no such thing as an average pilot. As Todd Rose explains, "One pilot might have a longer-than-average arm length, but a shorter-than-average leg length. Another pilot might have a big chest but small hips." Even when Daniels picked out just three of the ten dimensions of size—say, neck, thigh, and wrist circumference—less than 3.5 percent of pilots were average on all three dimensions.

How is this possible? On the surface, the finding that no airman conformed to the average seems not just confusing but contradictory. If you take a group of people and calculate the average of some trait, it surely must tell you something about the individuals within the group. After all, the average has been calculated from that sample of individuals. Yet focusing on averages can be misleading. Take the body length of weaver ants. There are two types of weaver ants—some very large, others very small. This means that if you take the average across all such ants, this average depicts none of the individual ants. The average is, in a certain sense, nonrepresentative. This kind of data is called a *multimodal distribution*.

Male height in humans, by contrast, conforms to a different type of distribution: the classic bell curve. This means that most people do indeed cluster around the average. There are many men who are five nine, but very few who are four nine or six nine. If you take a given person, you can be pretty confident he will be somewhat near the mean. But the design of a cockpit is not just about height. It is about human size across multiple dimensions. There is chest circumference, arm length, leg length, torso circumference, and so on. It is easy to assume that if you are high on one dimension (say, neck circumference) you will be high in another (waist size). But these correlations are, in fact, weak. This means that any metric that averages across these dimensions will obscure the diversity.

Take the two men in figure 9. If you took the average of all nine dimensions, they would be pretty much identical. And yet the man on the left is heavier and shorter and has a thicker neck, a shorter reach, a wider waist, and narrower shoulders, while the man on the left is taller, lighter, and so on. Summed across dimensions, they are both close to average. On any given dimension, however, they are often far from average.

We can see this point in a different way by taking another metric: IQ. It is easy to assume that if you take two people with an IQ of, say, 105, they will have similar scores on each of the various components of IQ, such as

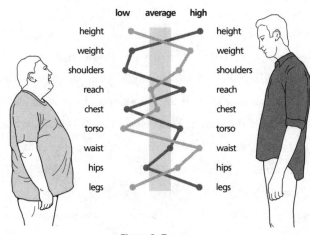

Figure 9. Two men

vocabulary, problem-solving, and so on. In fact, as Todd Rose has noted, the correlations are weak. You could be high on matrix reasoning, low on knowledge, medium on block design, high on symbol searching, and low on encoding—or vice versa. The single IQ metric is not expressing this variation, it is concealing it. And in most areas of performance, it is the variation that matters.

Daniels's work led to a dawning realization. A cabin standardized to the average pilot may sound logical, even scientific, but it is fraught with latent danger. The standardized cockpit was the root cause of the alarming incident rate, causing multiple crashes. And it forced the Air Force to think in a new way about design. Instead of requiring the pilot to conform to a standard cockpit, which suited almost nobody, they redesigned the cockpit to adapt to the diversity of individuals.

Sure enough, when planes were designed to enable pilots to vary the height of the seat, the distance of the joystick, and so on, the number of incidents plummeted. Moreover, the cost of creating this flexibility was minimal compared with the savings on incidents, not to mention the human cost of injuries and fatalities. In turn, the safety performance of the U.S. Air Force soared.

III

The standardized cockpit of the U.S. Air Force is not just a historic hazard, it is also a metaphor. It is just one example of the standardization of our world. We have standardized education, standardized working arrangements, standardized policies, standardized medicine, even standardized psychological theories. All, in their different ways, fail to consider human diversity. They treat all people as manifestations of a mythical average rather than as individuals. This takes us back to the point at the beginning of this chapter about how this flaw can cause us to overlook human diversity and lose its benefits. We are all different from one another. We have different physical dimensions, but also different cognitive traits, strengths and weaknesses, experiences and interests. Indeed, this is one of the most wonderful things about our species.

But if we differ in important ways, enlightened systems should, where possible, take account of this variation. Indeed, we should celebrate it. After all, how can we reap the benefits of human differences when we are crowbarred into rigid systems (and not just cockpits)? How can we harness diversity when we are deluded by averages that obscure the ways in which we differ from each other?

Let us take a brief example to highlight the logic. In an experiment at Google in 2014 a team of psychologists gave a short workshop to staff in sales and administration. Such work tends to be performed in a standardized way, at the same times, and with the same tools. This standardization is not physical, but conceptual. Indeed, the idea of building flexibility into such jobs seemed like madness. After all, these are the admin and salespeople, not the whizzy engineers.

But the workshop encouraged the professionals to think of their jobs not as fixed parameters, like inflexible cockpits, but as adjustable designs. They were taught to consider how they could play to their strengths, shaping the contours of their work around their interests and talents, as well as the objectives of the company. They were asked, in effect, to think of themselves as individuals with distinctive skills and insights rather than homogenous cogs in a machine. As Adam Grant, one of the coresearchers, put it:

> We introduced hundreds of employees to the notion that jobs are not static sculptures, but flexible building blocks. We gave them examples of people becoming the architects of their own jobs, customizing their tasks and relationships to better align with their interests, skills, and values—like an artistic salesperson

volunteering to design a new logo and an outgoing financial analyst communicating with clients using video chat instead of email. . . . They set out to create a new vision of their roles that was more ideal but still realistic.

What happened? Those who attended the workshop were rated by managers and coworkers as happier and higher performing, and were *70 percent more likely* to land a promotion or move into a preferred job when compared with a control group. Grant writes, "Instead of using only their existing talents, they took the initiative to develop new capabilities that enabled them to create an original, personalized job. They became happier and more effective—and qualified themselves for roles that were a better fit."

AVERAGES CAN BE VERY USEFUL, as we saw in chapter 2. You'll remember that the average forecast of six economists was significantly more accurate than the forecast of the top economist. Yet in this chapter we seem to be arguing that averages are bad. How can this make sense? Is there a contradiction in the analysis? In fact, these two perspectives are not just compatible but complementary.

The economic forecasters had different models. They expressed their estimates independently. They were free to come up with their own predictions. Taking the average of these different perspectives was a way of aggregating diverse information while filtering out the errors.

Standardization is different. It happens when people of different sizes are forced to inhabit the same cockpit. Or when people are asked to do their jobs in the same way, regardless of their differences, squeezing diversity out of the picture before it has even had a chance to manifest itself. It would be rather like forcing economists to use the same model, the model used by the average economist. This would effectively eliminate useful differences. Rebel ideas would dry up.

To put it another way, averaging diverse predictions is a way of exploiting diversity. Standardizing the way people work, or learn, or whatever else, risks squashing diversity. As Neil Lawrence, head of machine learning at Amazon, puts it, "When an average is being used well, it's harnessing the insights *from* multiple people. When it's used badly, it's imposing a solution *for* multiple people."

Standardization can be useful and valuable. With clothing, for example, off-the-shelf options may not always fit perfectly, but they enable consumers to gain the benefits of cheap, mass-produced apparel. Individualized solutions (made-to-measure) are typically more expensive, implying a trade-off between the bespoke and the generic. But often the generic is adopted not because it is more cost-effective but because we scarcely consider the alternative. That was the case with cockpits, where standardized designs were created not after a cost-benefit analysis but because few conceived of the possibility that a cockpit designed for the average pilot might not suit most pilots—at least until Lieutenant Daniels came long.

When institutions are too rigid, everyone suffers. This is true not just of organizations, but of science itself, when patterns of thought that orbit the concept of average cause us to overlook human diversity in subtle ways. With this in mind, let us return to Eran Segal. The risks of standardization extend far beyond what we put in our stomachs.

AFTER HIS LOW-CARB MARATHON RUN, Segal was finally zooming in on the flaw in dietary science. Dietary guidelines, like standardized cockpits, might seem rigorous, but they overlook a key variable: *the diversity of people.* Segal says:

A good example is the so-called glycemic index. This is a system of ranking foods according to how much they influence blood sugar. The way to obtain such an index is to take a group of people, get them to eat different foods, and then measure the response. This way you can obtain an index ranked from one to one hundred that rates food accordingly.

Described in this way, the glycemic index sounds like a gold standard of science. It is built around measurement and data. But it is also built around something else: the *average* response to food. What if people react to the same food in fundamentally different ways? People who base what they eat on the glycemic index might be eating the food that is, for them, unhealthy. In the spring of 2017, Segal and his fellow researchers conducted an experiment to test this possibility. The objective was to measure the response of subjects to two different types of bread: the commercially produced, white bread often demonized by the health lobby, and the handmade, whole-grain sourdough beloved by health nuts. As ever, the existing evidence was mixed. Some studies suggested that bread could reduce the risk of cancer, cardiovascular disease, and type 2 diabetes. Others suggested bread had little effect on health at all.

Segal's experiment was simplicity itself. He took a group of healthy people, none of whom were on any particular diet. They were then randomly assigned to two different groups. Some ate white bread every day for a week, while the others ate brown. Neither group was permitted to eat any other wheat products, and both were required to eat only bread for breakfast, while including it in other meals as desired. The two groups then took a two-week break, and then switched to the other diet.

Crucially, every person in the experiment was measured multiple times for how they were responding to the bread. Various measures were tracked, including inflammation response, nutrient absorption, and more.

Perhaps the most important measurement was the blood sugar response, which is crucial to health, and it is worth spending a moment or two to explain why. One of the first things biology students learn about the human body is the importance of glucose metabolism. After we eat, our body digests the carbohydrates, breaks them down into simple sugars, and releases them into the bloodstream. From that point, with the help of insulin, glucose is moved into the cells and liver, where it is used to synthesize glycogen for later use as energy.

However, insulin also signals cells to convert excess sugar into fat and store it—this is a primary reason for weight gain. Also, if too much glucose flows into the blood from food, it may cause an overproduction of insulin, pushing glucose levels too low. This makes us hungry and keen to eat more, even though we have had more than enough food. Sharp glucose spikes contribute to diabetes, obesity, cardiovascular disease, and other metabolic disorders. One study that followed two thousand people for more than thirty years found that higher glucose responses predicted higher mortality. Chronically high blood sugar levels put stress on your whole system. Steady blood sugar, on the other hand, with modest and gentle rises after eating, may reduce heart disease, cancer, and other chronic diseases such as excess fat and mortality. In short, the blood sugar response is important not just in terms of weight, but health, too.

When the results from Segal's bread experiment came rolling in, it turned out that the two different breads made no difference when it came to blood sugar response or any of the other clinical markers. Industrialized white bread and handmade sourdough had virtually the same effect. This seemed to imply that dietary advice should be neutral. If one bread is no better than the other, why not advise consumers to select the one that tastes better, or is cheaper?

And yet this "scientific" inference was based on the average response. What about the individual responses? Were people diverse in the way they responded? The results were remarkable. Some people showed benefits from eating the whole-grain sourdough and adverse effects from the commercial white, while others had the *opposite response*. Some showed little difference between the two, while others showed dramatic differences. "The whole data set was highly personal," Segal says. "You had to look at the individuals, not just the averages."

Why were the responses so different? Segal realized that just as multiple dimensions influence whether a person will fit into a cockpit, the human body has multiple dimensions that influence how an individual will respond to a given meal. These include intuitive variables such as age, genetics, lifestyle, and more. The body has multiple dimensions that determine a body's response.

Perhaps the most fascinating dimension is the microbiome, the bacteria we all host in our gastrointestinal systems. There are around forty trillion cells and up to a thousand different microbial species in our bodies. This universe has about two hundred times more genes than the host human and exerts a major influence on how we digest food, extract nutrients, and activate our immune system. And these microbiomes vary from person to person.

When you look at diet from this perspective, with different factors translating into different enzymes, genes, bacterial genes, and perhaps dozens of other unique factors, it seems almost absurd to suppose that any diet could be sensible for all, or even most, people. "The more I thought about it, the more curious it all seemed," Segal says. "Standardized dietary advice will always be flawed, because it only takes into account the food, and not the person eating it."

Segal's most ambitious study went further. Many experiments take a small group of people, give them a treatment or intervention, and then measure the average impact at a specific moment in time. Segal's experiment recruited almost a thousand subjects. Around half of these were overweight and a quarter were obese, matching the nondiabetic population of the developed world. These subjects were then connected to a glucose sensor and tracked every five minutes for an entire week, resulting in individualized blood sugar responses for almost fifty thousand meals.

The subjects logged everything they ate on a specially designed mobile app. They were allowed to eat what they wished, but they were required to eat a standardized breakfast: a rotating menu of plain bread, bread with butter, fructose powder mixed with water, or glucose mixed with water. This created a rich data set, including a total of 46,898 real-life meals and 5,107 standardized meals, with 10 million calories logged, along with associated health data. For an experiment involving nutrition, this was on a different scale from almost anything before. And instead of calculating an average response, Segal and his colleagues examined the response of every individual.

The results, when they came in, were stunning. For some people, eating ice cream led to a healthy blood sugar response, while sushi had the contrary effect. For others, it was the other way around. "For every medical or nutritional finding that came up, there were many people whose results were very different from it," Segal said. "Often people responded in diametrically opposed ways."

Keren—Segal's wife—was staggered by the results. A trained clinical dietician, she had seen dozens of patients in her clinic and relied on general guidelines to offer advice. With prediabetics, the advice is to stop eating ice cream and switch to complex carbs, like rice, instead. "I realized that I had been giving people advice that could harm them," she said. "It was sobering. Now I advise them to measure their own blood sugar responses. That way, they get a diet that works for them."

Talya is an archetypal example. A sixty-four-year-old retired pediatric nurse from northern Israel, she was heading toward diabetes. She was clinically obese and increasingly worried about her health. "I was putting on a lot of weight," she told me. "My blood sugar levels were very high." She was eating in a seemingly healthy way: omelet for breakfast, balanced meals through the day, and plenty of fresh fruit and vegetables. She grew her own produce in her backyard and particularly enjoyed apples and nectarines. "It seemed to be as good a diet as I could manage," she said. "I couldn't really figure out what I was doing wrong."

When she was given a glucose sensor so she could take regular measurements of her personal response to meals, she was dumbfounded. She spiked for nectarines, melon, and tomatoes. She also spiked for milk with 1 percent fat. Yet her blood sugar response was perfectly healthy for watermelon and 3 percent milk. "It was astounding," she said. "I had no idea this was happening."

Talya altered her diet to match up with personalized guidelines, losing almost 40 pounds and lowering her blood sugar levels by 20 percent. "No two people are exactly the same," she said. "We have different DNA, different biology. I am married to a man who is very skinny. Before, when we ate the same things, his blood sugar was fine. Now my blood sugar is coming down to normal. . . . Who on earth would have guessed that I would have a problem with nectarines!"

But Segal's study was not yet finished. The researchers then pulled all the data into an algorithm designed to predict blood sugar responses for new participants. Effectively, they were using an approach similar to the way online retailers like Amazon predict the kinds of books that shoppers will like. To test the algorithm, one hundred new people were recruited and then measured on personal characteristics such as blood, age, microbiome, and the like. This data was then fed into the algorithm. This was a significant test of the research. Would the algorithm more accurately predict how people would respond to different meals than standard carbohydrate counting?

The answer was an emphatic yes. "It was a huge thrill to see that we could take any person, even people who were not part of the original study, and predict their personalized glucose response to any meal with good accuracy," Segal said. "It gave us assurance that the algorithm was robust."

Finally, they recruited twenty-six new participants with prediabetes and asked the algorithm to design two diets for each subject. In the "good diet," the algorithm was asked to predict meals that would have a low blood sugar response. In the "bad diet," it was asked to predict meals with a high response.

By now, you won't be surprised to hear that the bad diet for some was similar to the good diet for others. One person's good diet was composed of eggs and bread, hummus and pita, edamame, vegetable noodles and tofu, and ice cream, while their bad diet was composed of muesli, sushi, marzipan candy, corn and nuts, chocolate, and coffee. As predicted, the bad diets were associated with abnormally high glucose levels and impaired glucose metabolism. On the good diet, despite the same number of calories, glucose levels remained completely normal, without a single spike across the entire week. "These results were frankly stunning to us," Segal said. "It was proof that you can manipulate your own blood sugar levels so significantly that you can go from prediabetic blood sugar levels to normal in one week, only by changing your food choices."

These results are important in their own right, but the key point—for our purposes—is that they deepen our comprehension of diversity. Presuming that pilots all conformed to the dimensions of the average pilot led to a litany of incidents in the 1940s–1950s. The same conceptual flaw has persisted, almost unnoticed, at the heart of nutritional science. Unless you consider the diversity of individuals, you are likely to design systems, guidelines, and much else that are defective or restricting or both.

When it comes to personalized nutrition, there is a long way to go. More studies are needed with longer-term follow-ups that measure health outcomes directly rather than via indirect indicators, such as blood sugar. More research is required to understand the microbiome, and more. The start, however, has been highly promising, and gives researchers a chance to overcome the contradictions that have bedeviled the field. Above all, it articulates a vital truth that science itself can dispose us to forget: diversity matters.

IV

In the spring of 2010, Michael Housman, a labor economist, was trying to figure out why some call center workers perform better than others. No matter how hard he looked, he couldn't find an answer. Nothing seemed to compute. He told me:

> I was working as chief analytics officer for a firm that sells software to employers to help them recruit and retain staff. We had data on fifty thousand people who had taken a forty-five-minute online job assessment and who were subsequently hired. We examined every aspect of the assessment to see if it held clues about longevity and performance. But we kept drawing a blank.

Housman's team had anticipated that those with a history of jumping around employers might, on average, leave quicker. They didn't. Staff could have had five jobs in previous years, or just one, but it didn't predict longevity at all. The team thought that certain aspects of personality revealed by the assessment might correlate with performance, too. That didn't stack up, either.

But then one of Housman's research assistants had a flash of insight. The team had data on the web browsers that had been used by the applicants to fill out the assessment forms. Some of the candidates had used Safari, others had used Firefox; some had used Internet Explorer, others

Chrome. Might the choice of web browser predict performance? To Hous-man, it seemed unlikely. Surely, this was just a matter of personal prefer-ence.

Yet the results were startling. Those who had filled out their assess-ments on Firefox or Chrome stayed in their jobs 15 percent longer than those who used Safari or Internet Explorer. They then checked the num-ber of absences from work. Again, they found the same gap. Those who used Firefox or Chrome had 19 percent fewer absences from work than those who had used Internet Explorer or Safari.

If this wasn't puzzling enough, the numbers related to performance were even more striking. Those who used Firefox and Chrome had higher productivity, higher sales, happier customers, and shorter call times. "It was one of the most emphatic sets of results we had found," Housman said. "These were big differences, and they were consistent."

What was going on? Housman said:

It took us a while to figure it out. The key is that Internet Explorer and Safari are preinstalled. PCs come with Explorer as part of the package, and Macs come with Safari. These are the defaults. To use them, you just need to turn on the computer. Chrome and Firefox are different. To use these pieces of software, you have to be curious enough to check if there are better options out there. Then, you have to download and install them.

It wasn't the software itself that was driving these differences in per-formance, it was what the choices revealed about differences in psychol-ogy. Some people have the tendency to accept the world as it is. They stick with the status quo. Others see the world as changeable. They wonder if there are better ways of doing things and, if so, act upon them. A seem-ingly inconsequential decision on which web browser to use revealed different positions on a psychological spectrum. Translated into the jobs they were doing, this meant many things. Remember, these were professionals working in call centers in retail and hospitality. Such jobs often have a set of scripts that are used to deal with consumer inquiries. It is easy to stick to a script. It represents the default. But every now and again, you meet a situation that isn't covered by the script, or in which a fresh approach might work better. Do you just stick to what you have always done? Or do you find a new way of solving a problem, or selling an idea, or pleasing the customer?

Those call center workers who could step outside convention performed significantly better. When the status quo wasn't good enough, they came up with something original. This mindset also helps explain why users of Chrome and Firefox stayed in their jobs longer and had fewer absences. Workers capable of altering the script are more likely to take action to fix problems, and make changes to their jobs that make them happier and more productive. Those who see the status quo as immutable are less likely to fix problems at work. They just put up with the default. Until they quit in frustration. "We were initially shocked by the size of the results," Housman said. "But we came to realize that the web browser decision shone a light on a crucial trait. The ability to question defaults makes a huge difference in a changing world."

The Housman experiment is rightly taken as evidence for the power of questioning the status quo, something that takes us back to chapter 4. These were people with an outsider mindset, capable of stepping outside the paradigm. This made them more productive and more fulfilled. They could solve problems, rather than simply endure them.

But there is another lesson, too. Think of the concept of best practice. One of the most familiar ideas in business, it hinges on a simple assertion: if there is a way of doing things that is proven to be superior, it makes sense for everyone to adopt it. In healthcare, for example, when doctors perform procedures in different ways, only some patients get the treatment with the best outcomes. Another aspect of best practice is also well understood. It is "best" not in an absolute sense, but in a comparative sense. It is the "best so far." If one can show that there is an alternative way of doing things that is superior to the status quo, then best practice should be amended. In this conception, best practice evolves through time, in a rational, data-led way.

But we should now see that this analysis, while useful, is also incomplete. Why? We can go back to the research of Eran Segal to expose the flaw. Suppose you ranked diets according to the blood sugar response for a population. You could conduct the study with rigor and determine that one diet was superior to all possible alternatives. And yet this *wouldn't be the optimum diet.* It would merely be the best standardized diet. A different approach would be to adopt personalization. This doesn't consist in comparing different diets across the whole population, but adopts flexibility at the level of the individual. And, as Segal showed, this approach can lower the blood sugar response, leaving an individual with a significantly healthier diet.

Now think of the call center workers. Many organizations test different scripts, compare the results, perform statistical tests, and come to what seems like a scientific conclusion about which script is the best overall. But this often misses the benefits of flexibility. The Housman experiment shows that when workers sensibly deviate from a script, they often perform better. This is partly because they are adapting to a new situation, as noted earlier. But it is also because it gives them a chance to play to their strengths, to bring their personalities into the conversation. The script is varying according to the individuality of the worker. To put it another way, best practice cannot be established by comparing standardized solutions; it also requires the comparison of different kinds of flexibility. And, given what we have learned about diversity, it is the more flexible systems that often win, whether one is talking about cockpits, diets, scripts, or anything else.

The question of how to build more flexibility into the world of work is now a major topic. It often centers around the scope to work from home, holidays, hours, and the like. Work is often more rewarding when professionals have the latitude to tailor schedules to personal commitments, and offering such flexibility means that organizations can access new talent (such as people who might not wish to work a standard 9 a.m.–5 p.m. job). This is particularly significant for Generation Y. According to one study, work-life balance was the single most important factor for young people in choosing an employer, something that is likely to become of growing importance.

And yet this represents only one facet of the power of flexibility. The deeper significance is expressed in diversity science and has not yet scratched the surface of the way institutions and societies function. This will prove to be a key source of innovation in the design of systems, allowing individuals to play to their strengths, and to bring their distinctiveness into the workplace.

Flexibility has dangers, of course. When we have the latitude to make changes, we also have the freedom to get things wrong. There is always a balance. But while institutions often focus on control and error, we rarely acknowledge the dangers of standardization that often lurk, undetected, at the heart of modern systems. In short, we need to become altogether more scientific about diversity.

V

Standardization is baked into our lives. Education reformers in the early nineteenth century designed schools with "standardized curricula, standardized textbooks, standardized grades, standardized holidays, and standardized diplomas." The idea was that education should not flex to the needs of individual learners, but that individuals should be crowbarred into the needs of the system.

The paradigm was mass production: schools would churn out students the way factories churn out widgets. They should be taught in the same way, at the same pace, with the same tools and the same textbooks, measured with the same tests. As Ellwood Cubberley noted in his influential 1916 guide, "Our schools are, in a sense, factories, in which the raw products (children) are to be shaped and fashioned into products to meet the various demands of life."

This approach had advantages over the disjointed system it preceded, but it also had limitations. After all, if youngsters are different from each other in important ways, flexibility should be designed in at the level of the student (a point wise teachers have always known). Indeed, there is good evidence to suggest that flexibility offers better results for students and schools. The 2015 PISA Tables showed that "adaptive instruction" was the second-most powerful predictor of high levels of educational outcome, rating above discipline, classroom size, and more. (The only factor more powerfully correlated with performance was recruitment of children from wealthy backgrounds.) Adaptive instruction is what you might expect—teachers who adapt to the needs of individual students, rather than getting everyone to do the same thing, at the same pace, at the same time.

A recent article by the author and teacher Maria Muuri summarizes the key tenets of Finnish education, often considered to be among the best in the world. Five of these factors dovetail with the themes of this book, including transversal skills, which seek to equip children with flexible thinking, and multidisciplinary learning, which helps students see how subjects are not separate silos but domains that can be bridged to forge new insights (recombination). Muuri also writes about why it is important for the system itself to be flexible. "Students are all individuals, so we can't teach them all in the same way," she writes. In Finnish schools, "there are usually at least five different levels of assignment in the same class at the same time. It also means that every student has their own specific goals." This is called *differentiation*.

Another key factor is diversity in students' assessment. She writes:

The new Finnish curriculum emphasizes diversity in assessment methods as well as assessment that guides and promotes learning. Information on each student's academic progress must be given to the student and guardians on a sufficiently frequent basis. . . . We set goals and discuss the learning process, and the evaluation is always based on the students' strengths.

Muuri also, rather wonderfully, writes about the way that students can benefit from cognitive diversity within study groups: "We make a point of having students from different backgrounds work together. I believe that there's always something that you can learn from someone who is different than you."

Some argue that there are parts of the educational system where personalization has gone too far. Others, such as Todd Rose, want it to go much further. This is a healthy debate, which should be guided by evidence. What is universally agreed is that providing for flexibility in otherwise-rigid systems can help all students flourish. Of course, the tyranny of average is about far more than education; it has infiltrated science more generally. A classic fallacy is to take averages based on male subjects and assume they apply to women, too. Think back to the cockpits. If these were ill designed for different sizes of men, imagine how much worse they would be for women, who are, on average, smaller. In her book *Invisible Women*, Caroline Criado Perez points out that piano keys were designed for the hand size of the average man, along with such things as police body armor and military equipment.

These physical design flaws are manifestations of a world of institutions designed for the average male, and these flaws invisibly make things harder for women. As Perez says, "One of the most important things to say about the gender data gap is that it is not generally malicious, or even deliberate. Quite the opposite. It is simply the product of a way of thinking that has been around for millennia and is therefore a kind of not thinking."

These conceptual confusions occur with the so-called hard sciences, too. A few years ago, Michael Miller, a neuroscientist at the University of California, Santa Barbara, conducted an experiment with twenty subjects who were placed in an fMRI machine. They were presented with a list of words, given a break, then presented with a list of new words. They were then asked to press a button every time they saw a word that had been on the original list. The brain scans of the subjects were then analyzed. The objective was to determine the neural circuits implicated in verbal memory. These are typically presented in a brain map, showing which part of the brain lights up, and will be familiar to anyone who has read a paper on neuroscience. What is perhaps less appreciated is that this map is calculated by averaging across subjects.

For some reason, Miller decided to look not at the average response but at the maps that detailed the individual responses. "It was pretty startling," he said in an interview with Todd Rose. "Most didn't look like the average map at all. . . . What was most surprising was that these differences in patterns were not subtle; they were extensive."

Think about that for a moment. Neuroscience, one of the most exciting branches of modern research, can provide misleading conclusions because the average brain map conceals the diversity of individual responses. Rose comments, "The extensive differences that Miller found in people's brains aren't limited to verbal memory. They've also been found in studies of everything from face perception and mental imagery to procedural learning and emotion." None of this means that neuroscience is flawed. Sometimes using averages makes sense. All too often, however, scientists use averages while scarcely conscious of doing so.[14] It is one step away from treating people not as diverse individuals, but as clones.

ONE OF MY FAVORITE EXPERIMENTS in diversity science was conducted by Craig Knight, a psychologist at Exeter University. Before becoming an academic, Knight was a salesman, traveling up and down the country. It was while standing in a large office in the West Midlands that he was struck by the dangers of standardization. He was looking down a vast row of melamine desks, all identical, stretching into the distance. The idea in vogue at the time was that all work areas should look alike and that workers should operate in standardized spaces. For Knight, it felt more than a little depressing. He tells me:

> It was called the lean office concept and it was all the rage at
> the turn of the millennium. The idea was that there should
> be no personal items. No photos. No paintings or plants. Such
> things were considered a distraction. If it could be scientifically
> proved that a particular type of work area was the most efficient,
> managers believed that everyone should conform to that.

As a traveling salesman, Knight noticed the lean concept in office after office, managers looking proudly at lines of standard spaces and workers beavering away in structured uniformity. Managers thought that they had hit on an empirical way of boosting productivity. Knight, who had a background in psychology, wasn't so sure:

---

[14] Formally, they take point averages as representative of a class of people while overlooking (or completely ignoring) the distribution from which it was calculated.

My hunch was that this had unintended consequences. If you put a gorilla or a lion in a lean enclosure, they are really miserable beasts. They get stressed, they fight, they become impotent, they die early. My suspicion was that humans are even more alienated by standardized spaces. People have their own personalities and characters, interests and ideas. I thought that people would want to create spaces for themselves.

It wasn't until Knight moved into academia a few years later that he had an opportunity to put his hypothesis to the test. His experiment, conducted with fellow researcher Alex Haslam, was ingenious. They took two groups of people and gave them tasks that would be typical in an office. Subjects had to check documents, process information, make judgments, and more.

The first group were put in the lean condition. They were all required to work in the same minimalist, superficially efficient space. In fact, almost precisely the conditions that Knight had noted while standing in that office in the West Midlands. The second group were also put into standardized workspaces, but with a difference. This time Knight put prints on the walls and plants by the desks. Knight called this the *enriched condition.*

What happened? Performance improved by 15 percent. Perhaps this isn't too much of a surprise. People work better in environments, even standardized ones, which are more human. "This confirmed what I had suspected for years," Knight told me. "Most people prefer an enriched environment. People said that the pictures and plants really cheered up the place. Lean spaces may work on an assembly line with regimented tasks, but not for cognitive or creative tasks."

Knight then took a third group and changed the setup once again. For this group, subjects were allowed to individualize their workspace. They could choose their own prints, their own plants, configuring the space to their own tastes, personalities, and preferences. "They were told to make themselves at home," Knight says. We might call this the personalized condition.

Now, from the outside looking in, many of the spaces in the personalized condition looked just like those in the lean and enriched conditions. After all, some people actually like a minimalist space, while others like an enriched space. On average, the latter was superior to the former when it was imposed upon subjects—but this was only on average. And this was the key point about the new condition: these new spaces were personally chosen. They were nongeneric. Like the cockpits with adjustable seats and pedals, and the customized diets created by Segal, these workspaces fitted around the dimensions of the particular worker.

The results, when they came back, were remarkable. Productivity soared. It was almost 30 percent higher than in the lean office condition— and 15 percent higher than the enriched condition. These are large effects. "Give people autonomy to create their own spaces, and they come up with something better than almost anything else you can give them," Knight says. "One participant said, 'That was smashing; I really enjoyed it. When can I move in?'"

The uplift in productivity can be divided into two components. The first is the autonomy element. People were choosing rather than being dictated to. They felt empowered, so were more motivated. This element is less to do with the choices and more to do with the act of choosing. But the second element was shaped not by the act of choosing, but by the power of personalization. People could design spaces that they liked. They could mold the space to their own characteristics. This may sound like a small thing, but it is actually a very big thing. It is an approach that takes diversity seriously.

VI

The pioneering work of Segal and his colleagues has been turned into a high-tech start-up called DayTwo. Although currently operating in a limited number of countries, the objective is to take the approach worldwide. The process is simple. You provide a stool sample and the result of a blood test. This enables the DayTwo lab to test your microbiome and to assess your blood sugar levels. This is then fed into the algorithm, allowing researchers to provide personalized food recommendations along with a searchable database of glucose predictions for one hundred thousand meals and drinks. This isn't as systematic as the experiment conducted in 2015, which measured blood sugar responses to every meal in addition to information on microbiome, but it nevertheless signals a direction of travel. Diet, like other branches of human science, is moving away from standardization and toward personalization.

Eric Topol, a professor of molecular medicine and one of the most respected medics in the world, was so intrigued by Segal's research that he volunteered to take a full test, tracking every meal and intake of liquid to determine blood-glucose response, as well as having his gut microbiome tested. Within weeks, he learned more about his own unique responses to food than would have been possible from trialing any number of standardized diets. He discovered not only that he has an unusual microbiome, but that he experienced severe blood-glucose spikes for food he had been eating for years. "My gut microbiome was densely populated by one particular bugger—*Bacteroides stercoris,* accounting for 27 percent of my co-inhabitants (compared with its average of less than 2 percent in the general population)," Topol wrote in the *New York Times.* "I had several glucose spikes as high as 160 milligrams per deciliter of blood (normal fasting glucose levels are less than 100 . . .)."

This discovery didn't just have implications for his health and longevity, but it also enabled him to make sense of the serial contradictions in dietary advice. "Despite decades of diet fads and government-issued food pyramids, we know surprisingly little about the science of nutrition. The studies have serially contradicted one another," he writes. "Now the central flaw in the whole premise is becoming clear: the idea that there is one optimal diet for all people."

In April 2019, scientists from DayTwo met with the senior leaders of the National Health Service in London. Further research is taking place, not just in Segal's lab but elsewhere, seeking to build more evidence. The goal is to use not merely the microbiome and genome to make dietary recommendations, but also other personal factors such as medication, sleep, and stress levels. Topol writes:

> What we really need to do is pull in multiple types of data . . .
> from multiple devices, like skin patches and smartwatches. With
> advanced algorithms, this is eminently doable. In the next few
> years, you could have a virtual health coach that is deep learning
> about your relevant health metrics and providing you with
> customized dietary recommendations.

Diet is merely one branch of this conceptual revolution. In almost all areas of our lives, we will find ourselves moving from the era of standardization to the era of personalization. If this transformation is guided with wisdom and data, it has the potential to improve health, happiness, and productivity, too. As Segal puts it, "Diversity is a part and parcel of humanity. It is time to take it seriously."

THE BIG
PICTURE

I

We have covered everything from the failures of the CIA to Rob Hall's heroism on the summit of Everest, and from the curious history of wheeled suitcases to the dangers of political echo chambers. We have seen that when it comes to innovation it is better to be social than smart, and how a fixation on averages can obscure individuality, a point that illuminates the deep flaws in dietary science, not to mention the alarming crashes of the U.S. Air Force in the late 1940s to early 1950s.

All of these examples and stories, experiments and conceptual explorations articulate the same underlying pattern. They reveal the power of diversity and the dangers of neglecting it. The success of organizations, and societies, depends on harnessing our differences in pursuit of our vital interests. When we do this well—with enlightened leadership, design, policy, and scientific insight—the payoffs are vast.

It is worth coming back, in this final chapter, to one of the biggest obstacles in this journey. We have called this the *clone fallacy*: thinking in a linear way about complex, multidimensional challenges. Obvious when stated, it nonetheless lurks across society. This is the obstacle, more than any other, that prevents people from pivoting from the individual to the holistic perspective.

Today, the major focus remains individualistic. We are preoccupied with helping individuals become smarter, more perceptive, more able to guard against biases. The fine works of Gary Klein and Daniel Kahneman are written from this standpoint. And yet while this perspective is important, we should never allow it to obscure the holistic perspective.

The organizing concepts in this book are holistic. The collective brain. The wisdom of crowds. Psychological safety. Recombinant innovation. Homophily. Network theory. The dangers of fine-grain assorting. The content of these concepts emerges not from the parts, but the whole. This is crucial in an era when our most pressing problems are too complex for individuals to solve on their own—an era where collective intelligence is moving front and center.

We will complete our journey into diversity science by widening the lens fully. Diversity doesn't just help explain the success of individuals and institutions; it sheds crucial light on the very evolution of our species. Our beginnings offer the ultimate contrast between individual and holistic perspectives, and also lay the clone fallacy to rest. Meanwhile, back in the present, diversity science transforms the way we live, work, and structure societies.

II

Our species dominates the planet. We thrive in virtually any habitat. If we include our domesticated animals, we account for 98 percent of the world's terrestrial vertebrate biomass. We create powerful technologies, theories, and art. We communicate with sophisticated languages. Our cousins, the chimpanzees, are confined to a small band of tropical African rain forest, but we obey no such constraints. As Kevin Laland, professor of behavioral and evolutionary biology at the University of St. Andrews, says, "Our range is unprecedented; we have colonized virtually every habitat on earth, from steaming rainforests to frozen tundra."

It raises the question: Why are humans so successful?

If you close the book and think about this for a few moments, you are likely to come up with an intuitive answer. Humans are intelligent. We have big brains. These enable us to solve problems that elude other animals. Our brains help us come up with new ideas, whether theories, technologies, or ways of communicating. This means that we are uniquely capable of subverting nature to our will. The basic idea is Big Brains lead to Great Ideas (i.e., technologies, culture, institutions).

This framework, which has long dominated our worldview, is not just wrong, but the inverse of the truth. It emerges from the individualistic perspective, placing the human brain at the center of the analysis. From a holistic perspective, the direction of causality is the other way around: Great Ideas lead to Big Brains.

This may sound odd, but tracing the argument will take our analysis of cognitive diversity to its zenith. Diversity, it turns out, not only drives the collective intelligence of human groups, but has driven the unique evolutionary trajectory of our species. Diversity, in a real sense, is the hidden engine of humanity.

To see why, consider that our ancestors had brains around the same size or slightly smaller than those of Neanderthals, a point that has been made by, among others, Joseph Henrich, professor of human evolutionary biology at Harvard University (whom we first met in chapter 3 with the theory of prestige). This implies that our ancestors may have been less intelligent than Neanderthals. As Henrich puts it, "In primates, the strongest predictor of cognitive abilities across species is brain size. Consequently, it is not implausible that we were dumber than the bigger-brained Neanderthals."

But our ancestors had a critical but often overlooked advantage: we were more social. We lived in larger, more densely connected groups. This was of seismic consequence, because a group of fellow animals nearby meant animals from which to potentially learn. Even if each member of this social group had only rudimentary ideas about such things as finding food, making tools, or whatever else, the density of such ideas meant that any one person—even a smart person—could learn more from the group than they could figure out in a lifetime on their own.

This meant, in turn, that natural selection started to favor good learners—that's to say, people skilled at observing what others were doing, and picking up ideas. These skills were not important for Neanderthals because they didn't have a sufficiently dense social group from which to learn. The point isn't that Neanderthals' ideas were inferior to those of our ancestors. Rather, learning from others is costly (time that could be spent hunting, for example), and there weren't enough ideas to repay the investment.

Once natural selection favored good learners, the trajectory of evolution itself shifted. For if people have the skill to learn ideas from the previous generation, and then add a few more to pass down to the next, ideas begin to *accumulate*. No single idea conceived by an early human was more sophisticated than any single idea conceived by a Neanderthal, but the overall corpus of knowledge was now growing—and recombining.

Among Neanderthals, innovations typically died when their creators died. Individuals were discovering new things, but they were not being shared across social groups or down generations. With the ancestors of humans, on the other hand, individual brains were being connected, within social groups and through time. Innovations were less likely to be lost. On the contrary, they were likely to be augmented. This is the dynamics of information spillover through evolutionary time.

To go back to the terminology of chapter 4, our ancestors were not smarter than Neanderthals by virtue of their individual brains. They were smarter by virtue of their collective brain. As Henrich says:

> Neanderthals, who had to adapt to the scattered resources of
> ice-age Europe and deal with dramatically changing ecological
> conditions, lived in small, widely scattered groups. . . . Meanwhile
> the African immigrants [our ancestors] lived in larger and more
> interconnected groups. . . . The extra edge created by more
> individual brainpower in Neanderthals would have been dwarfed
> by the power of social interconnectedness of the collective brain
> sizes of the Africans.

Think back to the difference between the geniuses and the networkers in the thought experiment in chapter 4. We noted that the geniuses were smarter than the networkers but less likely to be in possession of new innovations. Innovation is about the interplay between individuals and the networks they inhabit. As knowledge accumulates, it feeds back into the collective brain and into natural selection itself.

Indeed, this process has proved so powerful that the transition from a collection of individual brains to a true collective brain represents what biologists call a *major transition*. This is where a steep change in complexity is permitted by some alteration in the way information is stored and transmitted. Classic examples are the transition from prokaryotes (unicellular organisms) to eukaryotes (organisms whose cells have a nucleus enclosed within membranes) and from asexual clones to sexually reproducing populations.

The human collective brain represents our planet's most recent major transition, which has led to an accumulating body of ideas, but also a feedback loop that has altered genetic evolution itself. Why? Because the expanding corpus of ideas (sometimes called *cumulative culture*) created a selection pressure for bigger individual brains to store and categorize this rapidly growing body of information.

Over the last 5 million years, human brains grew from around 350 centimeters, comparable to a chimp, to 1,350 centimeters, with the bulk occurring in the last 2 million years. This expansion only hit the buffers 200,000 years ago due to constraints of the female birth canal, a key part of the primate body plan. If the baby's head grows too big, it can't get out (or is liable to kill the mother while trying to do so). This is why natural selection favored intense cortical folding, high-density interconnections, infant skulls that remain unfused in order to squeeze through the canal, and rapid postbirth expansion.

So, humans do indeed have impressively big brains, but note the direction of causality. The accumulation of ideas is driving the expansion of brains, not vice versa: Great Ideas (via accumulation and recombination of simple ideas) lead to Bigger Brains. As Laland puts it, "Once population size reached a critical threshold, such that small bands of hunter-gatherers were more likely to come into contact with each other and exchange goods and knowledge, then cultural information was less likely to get lost, and knowledge and skills could start to accumulate."

So, why didn't chimps and other animals join humans on this evolutionary pathway? Why is it only humans who have what biologists call *dual inheritance* (we inherit both genes and an ever-growing body of accumulated ideas)? The reason is that the emergence of collective brains faces a chicken-and-egg problem. We have already touched on the logic. Brains designed for learning from others are expensive. From an evolutionary perspective, such brains only make sense when there is already a decent body of ideas out there to acquire. And yet without the ability to learn from others, there won't be a sufficient volume of ideas in the local environment to justify this cost. This represents a fundamental constraint on the emergence of collective brains. Henrich calls it the *start-up problem*.

Gorillas, for example, could never justify this cost because they live in single-family groups with only one male and several females. Orangutans are solitary and do not pair-bond, which means that young orangutans often grow up with only their mother to learn from. Chimpanzees are more inclined to group living, but studies of infant and juvenile chimpanzees show that they only have access to their mothers as role models.

This is why these animals lack anything more than rudimentary technology. New innovations tend to die with their creators. They inherit genetic capacities, but they do not inherit a body of accumulating ideas. To the extent that Neanderthals were en route to true collective brains, they were outcompeted by the ancestors of modern humans when the latter left Africa. These other groups couldn't compete not because they were individually less intelligent, but because they were less collectively intelligent.

This perspective also helps explain the nature not merely of human brains but of human bodies. Once ideas became a stable part of the environment, they started to drive other aspects of genetic evolution. Take the invention of fire, one of the greatest rebel ideas in our species' history. We do not know who first managed to create fire, but we do know that humans were able to teach this skill to each other and to their children. That's to say, fire became part of the cultural ecology for early humans, passed down from generation to generation.

But this meant, in turn, that we didn't need such big guts to detoxify food. The food was already partially detoxified by the process of cooking. Natural selection, therefore, started to favor humans with smaller guts, freeing up metabolic energy required for the growth of our brains. We didn't need such large mouths and teeth, or such strong jaws, or colons, or intestines, all of which started to adapt to a culture with fire and cooking. Henrich writes:

> Techniques such as cooking actually increase the energy available from foods and make them easier to digest and detoxify. This effect allowed natural selection to save substantial amounts of energy by reducing our gut tissue. . . . This externalization of digestive functions by cultural evolution became one component in a suite of adjustments that permitted our species to build and run bigger brains.

Or take the fact that humans are among the greatest endurance runners on earth. We can hunt down antelopes, hartebeest, and the like, particularly in hot weather. Various traits enable us to do this, in particular our impressive capacity to perspire. We can sweat one to two liters per hour, which is a wonderful cooling mechanism.

But this raises a puzzle: our stomachs are too small to take in the large quantities of water required to sustain running over long distances. How, then, do we keep going with such inadequate storage? Why do we sweat so profusely when we cannot take on enough water in the first place? The puzzle is only solved when you factor in technology. Once we learned how to carry water in gourds, skins, and ostrich eggs, and this technology became a stable part of our environment, passed down the generations, we didn't need a large storage system in our bodies. We outsourced water storage just as we outsourced food detoxification, leading to a different evolutionary trajectory.

But note, once again, the direction of causality. Our efficient distance-running adaptations could not have evolved without the *prior* technology for external storage. As Henrich puts it, "The evolution of our complex, and rather extraordinary sweat-based thermoregulatory system could only happen after we have developed the idea of making water containers (and locating water sources)."

Ideas and technologies also modify our biology in a nongenetic way. The fact that you are reading these words means that you are literate. You have learned the technique of reading from your parents and teachers who, in turn, learned it from theirs. But in acquiring the ability to read, you also modified your brain. Learning to read changes the left ventral occipital temporal region of your brain, thickens the corpus callosum, modifies the superior temporal sulcus and inferior prefrontal cortex. This rewiring of the brain that occurs through the process of learning how to read is a biological modification associated with literate societies, but it is not a genetic modification. As Henrich says:

Reading is a cultural evolutionary product that actually rewires our brains to create . . . an almost magical ability to rapidly turn patterns of shapes into language. Most human societies have not had a writing system, and until the last few hundred years, most people did not know how to read or write. This means that most people in modern societies (those with high reading proficiency) . . . have different brains with somewhat different cognitive abilities than most people in most societies across human history. . . . The crucial point is that cultural differences are biological differences but not genetic differences.

Rebel ideas have shaped our brains, our bodies, our social norms, and our institutions. They have also shaped our psychology. For once we could learn more from our social group than we could ever learn in a lifetime on our own, natural selection started to favor those who were skilled at extracting ideas from other brains. This meant the ability to pay attention to those from whom we had the most to learn. As Henrich puts it:

> Once [ideas] began to accumulate . . . the main selection pressure on genes revolved around improving our psychological abilities to acquire, store, process, and organize the array of fitness-enhancing skills and practices that increasingly became available in the minds of the others in one's group. . . . This process can be described as autocatalytic, meaning that it produces the fuel that propels it.

III

This whirlwind tour of human evolution provides us with the ultimate contrast between individual and holistic perspectives. The human brain is impressive, but the success of our species is in the intricate web of connections, stretching across the planet and back through history, that has led to the evolution of a vast body of ideas, technologies, and culture. For an estimated two million years this has guided the process of human genetic evolution, expanding the brain, altering our physiology, and feeding back into the body of ideas.

Our species is, in this sense, *constructed on diversity.* We are unique in the way that different ideas, experiences, lucky discoveries, and recombinations sweep through our social networks, building the collective brain, expanding collective intelligence, and altering the trajectory of natural selection. It is the diversity of these ideas that makes us smart. Stripped of the cumulative body of ideas, the naked human brain is far less impressive.

Anthropologists have measured our naked cognitive capacity (i.e., minus access to cumulative information) by comparing human babies with similarly aged chimps and other primates. At this stage of development, human toddlers have absorbed quite a bit of information from their parents, but far less than children of ten or even five. A study by researchers at the Max Planck Institute for Evolutionary Anthropology in Leipzig pitted human toddlers (two and a half years old) against chimpanzees and orangutans in tests of spatial memory (subjects had to recall the location of an object), causality (assessing shapes and sounds), and other cognitive tasks. The results were pretty much identical across all challenges. The humans and chimps performed at about the same level, with the orangutans slightly behind.

The one task that the humans excelled at was social learning. This involved observing a demonstrator use a complex technique to extract food from a narrow tube. The human children picked up the technique right away and were able to apply it instantly. The other primates were unable to make sense of what they had seen, or put it into practice. Henrich writes:

> On the social learning subtest, most of the two-and-a-half-year-old humans scored 100 percent on the test, whereas most of the apes scored 0 percent. Overall, these findings suggest that the only exceptional cognitive ability possessed by young children in comparison with two other great apes relate to social learning and not to space, quantities, or causality.

Let's acknowledge how odd this sounds. Humans are smart because we have evolved to connect with other brains. This is why a human child by the age of nine or ten can beat any other primate of any age on pretty much any of the cognitive tasks in the experiment. The body of knowledge absorbed from adults equips the brain with outsize power.

Chimps and orangutans do not improve as they get older. Once their brains have reached maturity at age three, that is as good as they will ever get. There is no collective brain to plug into, no corpus of ideas, no cumulative culture. And, even if there were, they have not developed the mental capacities to extract this information from other animals because in their evolutionary history there was no selection pressure to build such capacities. Michael Muthukrishna sums it up: "Why is it that humans are so different from other animals? It's not because of our hardware. It's not because we have these giant brains that make us more intelligent than other animals. In fact, some chimpanzees can beat us on basic working memory tasks. . . . What makes us different from other animals is our collective brain." Laland writes, "Humanity's success is sometimes credited to our cleverness, but [ideas] are actually what make us smart. Intelligence is not irrelevant, of course, but what singles out our species is an ability to pool our insights and knowledge, and build on each other's solutions."

This analysis may seem a little disparaging to the individual human brain. After all, it is considered the most complex entity in the known universe. We are proud of our cognition and processing. But the underlying point can be made using the brain itself as a metaphor. The brain is composed of innumerable neurons and axons. It has a complex system of many parts. And yet the intelligence of any given brain does not emerge from the intelligence of its parts. No single neuron is responsible for coming up with an insight. Rather, the brain's power is based on their interaction. As Marvin Minsky writes in *The Society of Mind*, "You can build a mind from many little parts, each mindless by itself."

The individual brain stands in relation to the collective brain somewhat as the neuron stands in relation to the individual brain. The metaphor isn't precise since individual brains do come up with insights on their own. Unlike neurons, they are not dumb. But the conceptual point stands: whether we are operating in normal time (measured in minutes, hours, years, and centuries) or evolutionary time (measured in hundreds of thousands of years), human development relies on the way diverse brains interact far more than on the constituent brains themselves.

We are left with a scintillating vision. Our species is the most formidable on the planet not because we are individually formidable, but because we are collectively diverse. By bringing different insights together, by connecting within and across generations, by recombining rebel ideas, we have created breathtaking innovations. It is our sociality that drove our smartness, not the other way around.

Cognitive diversity is not merely the ingredient that drives the collective intelligence of human groups, it is also what has driven the ascendance of our species, creating the latest of the major transitions on earth. This is, in many ways, the ultimate testament to the power of diversity.

HAVING LOOKED AT THE BIG PICTURE, let's narrow the focus for a final time to put the lessons of this book into practice. How can we harness diversity in our jobs and in our lives? The following applications all have immediate relevance to how we live, work, and structure societies.

Unconscious Bias

Unconscious bias, which dominates many contemporary debates on diversity, refers to the way that people are denied opportunities not because of a lack of talent or potential but because of arbitrary factors such as race or gender.

Perhaps the most intuitive example of unconscious bias emerged in the 1970s. At that point, orchestras in the United States (and elsewhere) were dominated by men. The reason is simple: those who conducted the auditions thought that men were, typically, better musicians. This was a meritocracy, they insisted. Men were said to be more accomplished pianists, violinists, and so on.

But Claudia Goldin of Harvard and Cecilia Rouse of Princeton had an idea: Why not conduct auditions behind screens? This meant that the selection panels could hear the music and assess its quality, but could not see the gender of the musician playing it. When these screens were introduced, women's chances of making it through the first round increased by 50 percent, and in the final rounds by 300 percent. The number of female players in major orchestras has since increased from 5 percent to nearly 40 percent.

What is fascinating is that the recruiters didn't realize they were discriminating against women until the screens were introduced. Only then could they see that they had been assessing candidates not merely on skill, but through the filter of stereotypes about what a good musician ought to look like. Eliminating bias was good not just for female musicians, but also for orchestras. They were recruiting talent regardless of what it looked like.

Unconscious bias tends not to manifest itself when the differences between candidates are obvious. After all, why would an employer deliberately choose an inferior performer? This would harm the organization itself. It is only when candidates are similar in ability, when the recruiter has what psychologists call *discretionary space*, that unconscious bias takes on greater significance.

Take a study in which university students were given the task of deciding between candidates for a job. When Black candidates were clearly superior to white candidates, they were almost always selected. The same was true of the white candidates. It was only when the CVs were similar in quality that unconscious bias kicked in. At this point, students showed a small but significant tendency to go with white candidates. They did not do so consciously. They were surprised when the bias was pointed out to them. Had any given decision been challenged in court, discrimination would have been nigh impossible to prove. And yet the stereotype that Black people are of lower ability than whites influenced the way they unconsciously processed the CVs.

The cumulative consequences of these small biases are significant. To make it anywhere in life requires a series of sliding door evaluations: you have to gain selection to the school debating team, to the summer internship, to university, to secure your first job, to gain promotion, and so on. These are merely the headline examples. Evaluations are baked into almost all of our daily interactions. Now think of the mathematics of such a sequence. Just a 10 percent bias against Black people in each of ten evaluations reduces their probability of getting to the top by a massive 90 percent, a point made by Scott Page.

Consider, too, how this creates perverse incentives. To attain credentials in the first place requires hard work and sacrifice, not just at school and university but beyond. Success requires a willingness to defer gratification in myriad ways. But if the payoff associated with these credentials is so grievously diluted, why would one bother to do the hard yards? Roland Fryer, a Harvard economist, has shown just how distorted the payoffs to education can become for minority groups. And this hints at what has become known as *structural bias*: the way that the legacy of historical injustice, unconscious discrimination, and skewed incentives can harden into concrete barriers for certain sections of the population.

Dismantling unconscious bias, then, is not just a powerful first step in creating a fairer system; it is also a first step in creating a more collectively intelligent society. It gives people from all backgrounds a chance to pursue their talents, broadening the body of people with the knowledge to contribute to our most pressing challenges. Combating structural discrimination should be near the top of any political agenda.

This takes us back to the screens in the orchestra auditions. These were effective because they took the subjectivity out of recruiting. Biases were removed through the design of the evaluation itself. The presence of the screens had deeper effects, too, giving aspiring female musicians confidence that their talent would be fairly evaluated—thus conferring a bigger incentive to gain the qualifications in the first place.

In her book *What Works*, Harvard academic Iris Bohnet offers an extensive analysis of different measures to bolster the objectivity of evaluations. These include "blinding" CVs (by removing certain types of demographic information) and altering the way companies search for new recruits, how they advertise positions, where they post job openings, how they evaluate applicants, how they create short lists, how they interview candidates, and how they make their final selections.

Eliminating unconscious bias is a vital technique when it comes to enlarging collective intelligence, but it is by no means sufficient. Think back to Bletchley Park. Suppose that the recruiters had been keen to hire mathematicians. By removing unconscious bias, they would have been in a position to identify the top mathematicians, uncorrupted by stereotypes.

Yet this would have not helped them find crossword experts and other idiosyncratic minds that turned out to be so crucial to cracking the Enigma code. Eliminating unconscious bias assists with finding the best individuals regardless of race or gender—but it doesn't, in and of itself, optimize cognitive diversity. These two challenges are conceptually distinct. Great organizations need to do both.

Shadow Boards

Another exciting way that cutting-edge companies are harnessing diversity is through the use of "shadow boards." These consist of young people who advise executives on key decisions and strategies, thus lifting the conceptual blinkers that can attach to age. Each of us grew up at a particular time and absorbed a particular cultural and intellectual paradigm. This influences the way we think in so many ways that we can become unconscious of it.

This applies as much to science as to anything else, a point that has been made by the philosopher Thomas Kuhn. Scientists often operate according to assumptions and implicit theories, which can place constraints on the development of new insights and ideas. This is what prompted the great physicist Max Planck to say that "science advances one death at a time."

Shadow boards typically consist of a group of the most able young people, drawn from across an organization, who have regular input into high-level decision making. This enables executives to "leverage the younger groups' insights and to diversify the perspectives that executives are exposed to," as management experts Jennifer Jordan and Michael Sorell wrote in the *Harvard Business Review* in 2019. This, in turn, drives a greater flow of rebel ideas.

Anyone who has struggled with a new technology, and been amazed at how quickly younger people take to it, will grasp the significance of a shadow board. Anyone who has been struck by how differently young people think about old problems will also grasp the logic. It should, then, be no surprise that companies with well-integrated shadow boards have reaped vast rewards.

Jordan and Sorell contrast the different fortunes of Prada and Gucci, two high-end fashion brands. Prada has traditionally enjoyed high margins but experienced a slump between 2014 and 2017. Why? A public announcement in 2018 acknowledged that the company had been "slow in realizing the importance of digital channels and the blogging online 'influencers' that are disrupting the industry." CEO Patrizio Bertelli said, "We made a mistake."

As for Gucci, they created a shadow board of young people who had consistent interaction with the senior team. "They talk through the issues that the executive committee is focused on, and their insights have 'served as a wake-up call for the executives,'" Jordan and Sorell write, quoting Gucci CEO Marco Bizzarri. Gucci's sales have since grown 136 percent—from 3,497 million euro (FY 2014) to 8,285 million euro (FY 2018)—a growth driven largely by the success of both its internet and digital strategies. In the same period, Prada's sales have dropped by 11.5 percent, from 3,551 million euro (FY 2014) to 3,142 million euro (FY 2018).

A Giving Attitude

Successful collaboration requires a particular attitude. One has to be willing to offer one's insights to others, to share one's perspective, to impart one's wisdom. It is only by giving that we gain the opportunity, in turn, to receive. In fact, perhaps the most powerful evidence for the growing importance of diversity is that people with a giving attitude are becoming ever more successful.

Consider a study of more than six hundred medical students, which found that the individualists—those who focused on their own progress, and cared little for others—performed very well in their first year. These "takers" were good at extracting information from those around them, and by offering little in return they were able to focus on their own progress. Those who were more generous with their time and were willing to offer insights to their fellow students, the "givers," got left behind.

But here is the curious thing. By the second year, the more collaborative cohort had caught up, and by the third year they had overtaken their peers. By the final year, the givers had gained significantly higher grades. Indeed, the collaborative mindset was a more powerful predictor of school grades than the effect of smoking on lung cancer rates.

What was going on? The givers hadn't changed, but the structure of the program had shifted. Adam Grant writes in his book *Give and Take*:

> As students progress through medical school, they move from independent classes into clinical rotations, internships, and patient care. The further they advance, the more their success depends on teamwork and service. Whereas takers sometimes win in independent roles where performance is only about individual results, givers thrive in interdependent roles where collaboration matters. As the structure of class work shifts in medical school, the givers benefit from their tendencies to collaborate more effectively.

This is a finding that keeps re-emerging across the social sciences: people with a giving approach are flourishing. This is not a hard-and-fast rule; we can all think of people who have a taking attitude, who hate sharing credit, but who have nevertheless achieved marvelous things. The world rarely fits into neat categories. But the evidence suggests a broad pattern in favor of a giving approach. It also shows that the most successful givers are strategic, seeking out meaningful diversity, and cutting off collaborations if they are being exploited. This enables them to benefit from the upside of successful teamwork, while reducing the downside of partners who free ride. As one researcher put it, "The giver attitude is a powerful asset when allied to social intelligence."

This willingness to give, to collaborate, has longer-term effects, too. You can see this in an experiment led by Professor Daniel Levin of Rutgers Business School, in which he asked more than two hundred executives to reactivate contacts that had been dormant for at least three years. The subjects asked two of these contacts for advice on an ongoing project at work.

They were then asked to rate the value of this advice when compared with reaching out to two people on the same project.

Which contacts provided fresher insights, better ideas, stronger solutions? The answer was clear. The dormant ties offered significantly higher-value advice. Why? Precisely because they were dormant, these contacts were not operating in the same circles, or hearing the same stories, or having the same experiences. The dormant ties were offering a diversity bonus—and it counted for a lot.

People who give construct more diverse networks. They have more dormant ties. They have access to a greater number of rebel ideas. By giving in the past, givers enjoy greater scope to reach out for ideas when it matters. As one executive put it, "Before contacting them, I thought that they would not have too much to provide beyond what I had already thought, but I was proved wrong. I was very surprised by the fresh ideas."

The willingness to share, to offer knowledge and insight and creative ideas, pays huge dividends in a world of complexity. It is the glue of effective collaboration, not just in the moment, but over time. The benefits compound. As Adam Grant writes, "According to conventional wisdom, highly successful people have three things in common: motivation, ability, and opportunity . . . [but] there is a fourth ingredient: success depends heavily on how we approach our interactions with other people. Do we try to claim as much value [for ourselves] as we can, or do we contribute value . . . ? It turns out that these choices have staggering consequences for success."

IV

Today, we stand on the brink of a revolution: a revolution triggered by a new way of thinking about diversity. Diversity is often regarded as a politically correct distraction, an issue of morality and social justice, but not of performance and innovation. It is debated in vague terms, people talking past each other. Our conception of diversity is not just incomplete but often radically defective.

This is why people are surprised that the average prediction of six economists beats the prediction of the most accurate economist, or that a group of perceptive but similar individuals is collectively myopic. It is why many scientists are surprised to discover that today's hit papers are recombinant, and why it seems odd that ideas drove the expansion of the human brain rather than the other way around.

It is only when we absorb the truths of diversity science that our perspective shifts. We begin to see intelligence as built not merely on the intellectual brilliance of individuals, but also on their holistic diversity. We see that innovation is about not merely the insights of specific people, but also the networks that permit their recombination. And we see that the success of humanity is less about individual brains than the emergent properties of the collective brain.

These truths have practical implications. Think back to our discussion of homophily, which operates like an invisible gravitational force, pulling teams and institutions toward homogeneity. We unconsciously enjoy being surrounded by people who think in the same way, who share our perspectives, who corroborate our prejudices. It is comforting and validating. It makes us feel individually smart even as we are becoming ever more collectively stupid.

But could there be any more powerful way to combat homophily than through an understanding of diversity science? Why would we wish to surround ourselves with like-minded people when this undermines the objectives of the group? Why would we enjoy the experience of having our views serially corroborated when it means that we are learning nothing new? Why would we crave cultures of conformity when this silences the rebel ideas that catalyze innovation?

The very meaning of collaboration changes when we think about diversity in a new way. Honest dissent is not disruptive, but imperative. Divergent opinions are not a threat to social cohesion but a contribution to social dynamism. Reaching out to outsiders for new ideas is not an act of disloyalty but the most enlightened form of solidarity. Without the innovation driven by recombination, how can any group keep pace with a fast-changing world?

To put it another way, you can only build a *culture* of diversity when you have first grasped the *concepts* of diversity. Bridgewater, one of the world's top hedge funds, grounds new recruits in the holistic perspective. They read a set of principles that, among other things, articulate diversity science. This is why people are considered loyal not when they agree, parrot, and validate, but when they honestly disagree, challenge, and diverge. They are applauded not when they stay rigidly within institutional boundaries but when they seek new ideas. As Ray Dalio, the leader of Bridgewater, says:

> Great cultures bring problems and disagreements to the surface and solve them well, and they love imagining and building great things that haven't been built before. Doing that sustains their evolution. In our case, we do that by having an idea meritocracy that strives for meaningful work and meaningful relationships through radical truth and radical transparency.

What is true at the level of organizations is also true at the level of societies. Cultures that encourage new ideas, foster dissent, and have strong networks through which rebel ideas can flow, innovate faster than those held back by cultures of intellectual conformity. As Henrich has put it:

> Once we understand the importance of collective brains, we begin to see why modern societies vary in their innovativeness.

It's not the smartness of the individuals. . . . It's the willingness and ability of large numbers of individuals at the knowledge frontier to freely interact, exchange views, disagree, learn from each other, build collaborations, trust strangers, and be wrong. Innovation does not take a genius or a village; it takes a big network of freely interacting minds.

These insights have been associated with radical thinkers and philosophers since at least the time of the Ancient Greeks, but they are today supported by formal theories and extensive data. The contribution of diversity to the dynamism of societies has, in this sense, moved from the terrain of intuition to hard science. Diversity is the critical ingredient that can help us solve our most pressing problems, from climate change to poverty, and help us break free of the echo chambers that are coming to disfigure our world. John Stuart Mill, the nineteenth-century English philosopher who stands as one of the most eloquent advocates for diversity in history, said:

It is hardly possible to overrate the value, in the present low state of human improvement, of placing human beings in contact with persons dissimilar to themselves, and with modes of thought and action unlike those with which they are familiar. . . . Such communication has always been, and is peculiarly in the present age, one of the primary sources of progress.

V

Let us finish this book where we started. In the years after 9/11, the CIA started to wake up to its crippling homogeneity. One sign of this awakening was the hiring of Yaya Fanusie, an African American Muslim who grew up on the West Coast and graduated with a degree in economics from the University of California, Berkeley, before winning a Fulbright scholarship and securing a postgraduate degree from Columbia in international affairs, international finance, and business. He converted to Islam in his early twenties, becoming a devout believer. I interviewed Fanusie on an early spring morning about his experiences at the CIA. He said:

> When I joined the agency in 2005, I was initially staffed in economic analysis, not counterterrorism. In some ways, that made sense given my background in economics. Just because I am a Muslim doesn't automatically mean that I should work on jihadist counterterrorism. But I came to feel that I might be able to provide unique insights, given my background. After the July 7 bombings in London, I asked to be transferred into the section combating Al Qaeda.

Fanusie quickly made his mark. After a briefing in the White House Situation Room, he became suspicious about Anwar al-Awlaki, an American Muslim preacher born in New Mexico to Yemeni parents. Between the mid-1990s and 2001, al-Awlaki had been an imam variously in Denver, San Diego, and Northern Virginia. Some of the 9/11 bombers had worshipped at his mosques. After he left the United States in 2002, he went first to the United Kingdom, then to Yemen. His sermons became ever more extremist. Fanusie says, "He was a great storyteller, interweaving American English with classical Arabic. His lectures were often hours in length. He was implicated in a kidnapping plot in 2006, at almost precisely the moment that I moved into counterterrorism. It was clear that he was focused on reaching young Western Muslims in particular." Fanusie conducted an exhaustive survey of his historical sermons, spotting clear warning signs. He said:

Al-Awlaki laid out an argument for Muslims to join the jihadist cause. These were not random musings, as some had supposed, but strategic directives. I could see the way that he was taking aspects of teaching, and cleverly molding them to the psyche of the millennial Western Muslim. After his release from prison, he started his own blog. He was in full recruitment mode, pulling in young Muslims from the United States, Europe, and elsewhere to Yemen. He literally weaponized them.

One Nigerian follower set his underwear on fire over Detroit one Christmas Eve in a failed airline plot. It wasn't fully understood by the intel community how one army major named Nidal Malik Hasan, conflicted and self-tortured by his role in the U.S. military, sought out al-Awlaki for advice. And when Major Hasan opened fire on his fellow servicemen and -women at Fort Hood, killing thirteen and injuring dozens, al-Awlaki posted on his blog that Nidal Malik Hasan had done the right thing.

Fanusie, a devout Muslim and American patriot, continued to analyze al-Awlaki's teachings, gradually becoming aware of the threat he posed to the West. "He was telegraphing his passes through his blogging and media interviews. . . . But you had to know what to look for. The key thing was to see what was happening, to see what he was building toward, so that we were in a better position to thwart his aspirations."

In April 2010, al-Awlaki was placed on a CIA kill list by President Barack Obama. On September 30, 2011, he was eliminated while in hiding in southeast Yemen in a strike carried out by Joint Special Operations Command, under the direction of the CIA. By this time, he was believed by the U.S. government to be one of the most dangerous figures in the world, described by one Saudi radio station as the "bin Laden of the Internet."

In 2015, Scott Shane, national security reporter for the *New York Times* who wrote a book on al-Awlaki called *Objective Troy: A Terrorist, a President, and the Rise of the Drone,* said:

> He was by far the most popular, most influential English-
> language recruiter for Al Qaeda and for the whole jihadist
> cause. . . . He just became one of the most powerful and effective
> voices persuading people to join Al Qaeda. . . . He sort of pioneered
> a do-it-yourself approach in a very explicit way. . . . If you need
> to know how to build a bomb, [he had] articles on that. In some
> ways, he pioneered what we're now seeing from the Islamic State,
> from ISIS, in terms of encouraging people in the West not to wait
> for instructions but to just go ahead and come up with an attack.

I ask Fanusie about the case for diversity in intelligence. He says:

> It is often said in intelligence circles that too few minority
> candidates apply to the CIA. Also, when you have candidates
> with significant foreign national (non-U.S. citizen) connections
> in their family background, there will often be counterespionage
> concerns that impact the hiring process. Recruiters tend to recruit
> candidates whose background resonates with them, usually
> because of common experiences, culture, and outlook. It is no
> coincidence that I was recruited by a Black woman.

"Is there a risk that hiring diverse candidates might dilute the quality
of the CIA?" I ask. Fanusie replies:

> You should never hire people just because of their cultural or
> ethnic background. That would be a dangerous mistake. But
> when you widen the net of recruitment, you also broaden the pool

of talent. It gives you the chance to hire outstanding people who are also diverse. And this has knock-on consequences. With more high-class people from minority backgrounds, it encourages new people to apply, broadening the pool still further.

The CIA has made strides toward meaningful diversity since 9/11, but the issue continues to dog the agency. An internal report in 2015 was damning about the lack of diversity in senior positions. As John Brennan, then director, said, "The study group took a hard look at our agency and reached an unequivocal conclusion: CIA simply must do more to develop the diverse and inclusive leadership environment that our values require and that our mission demands."

As for Fanusie, he is now senior fellow for the Center on Economic and Financial Power at the Foundation for Defense of Democracies. He is a leading thinker on intelligence, a regular speaker at international conferences, and the founder of a podcast that moves beautifully between his own journey as a Black American Muslim and some of the most pressing issues of global security. When he left the CIA in 2012, he received a plaque signed by Michael Leiter, former director of the National Counterterrorism Center. It thanked Fanusie with the words "You regularly had impact at the highest levels of the U.S. government."

CROSSWORD SOLUTION

T	R	O	U	P	E	█	S	H	O	R	T	C	U	T
I	█	L	█	S	█	█	█	O	█	E	█	U	█	A
P	R	I	V	E	T	█	A	R	O	M	A	T	I	C
S	█	V	█	U	█	A	█	D	█	I	█	T	█	K
T	R	E	A	D	█	G	R	E	A	T	D	E	A	L
A	█	O	█	O	W	E	█	█	D	█	█	R	█	E
F	E	I	G	N	█	N	E	W	A	R	K	█	█	█
F	█	L	█	Y	█	D	█	R	█	I	█	T	█	S
█	█	█	I	M	P	A	L	E	█	G	U	I	S	E
S	█	E	█	█	E	█	█	A	S	H	█	N	█	N
C	E	N	T	R	E	B	I	T	█	T	O	K	E	N
A	█	A	█	O	█	O	█	H	█	N	█	L	█	I
L	A	M	E	D	O	G	S	█	R	A	C	I	N	G
E	█	E	█	I	█	I	█	█	█	I	█	N	█	H
S	I	L	E	N	C	E	R	█	A	L	I	G	H	T

Crossword 5,062
Daily Telegraph, January 13, 1942

ACKNOWLEDGMENTS

As a kid who grew up with a dad who was born and raised in Pakistan, and a mom from North Wales, diversity has been part and parcel of my life. The idea for this book started to form when I realized that diversity is not just about race or cultural background but has implications that span everything from business to politics and from history to evolutionary biology. I am hugely grateful to a wonderfully diverse group of people who read early drafts of this book and offered suggestions. These include Adil Ispahani, Leona Powell, Neil Lawrence, David Papineau, Michael Muthukrishna, Kathy Weeks, Andy Kidd, Priyanka Rai Jaiswal, and Dilys Syed.

I would also like to thank my superb editor Nick Davies, and agent Jonny Geller. I have had terrific support from my colleagues at the *Sunday Times*, which is a wonderful publication to work for. I am particularly grateful to Tim Hallissey, my editor for more than fifteen years.

The intellectual inspirations for the book are, as you'd expect, diverse, but I would like to acknowledge two thinkers, in particular. The work of Joseph Henrich, professor of human evolutionary biology at Harvard University, has influenced the book in multiple ways, as has that of Scott Page, professor of complex systems, political science, and economics at the University of Michigan, Ann Arbor. I would like to thank both for reading drafts and taking the time to discuss the central issues.

One of the most wonderful things about writing a book of this kind is coming into contact with a huge variety of books, research papers, and case studies. I have tried to reference all of these in the endnotes for anyone wishing to look at specific topics in more depth, but these are some of the books that have been particularly influential: *The Secret of Our Success* by Joseph Henrich, *The Difference* and *The Diversity Bonus* by Scott Page, *Constructing Cassandra* by Milo Jones and Philippe Silberzahn, *The End of Average* by Todd Rose, *Into Thin Air* by Jon Krakauer, *The Secrets of Station X* by Michael Smith, *Regional Advantage* by AnnaLee Saxenian, *Echo Chamber* by Kathleen Hall Jamieson and Joseph N. Cappella, *Friend and Foe* by Adam Galinsky and Maurice Schweitzer, *Invisible Women* by Caroline Criado Perez, *The Blunders of Our Governments* by Anthony King and Ivor Crewe, *Der Spiegel's Inside 9/11*, *Infotopia* by Cass Sunstein, *The Righteous Mind* by Jonathan Haidt, *What Works* by Iris Bohnet, *Give and Take* by Adam Grant, *Principles* by Ray Dalio, *The Origins of Political Order* by Francis Fukuyama, *The Rational Optimist* by Matt Ridley, *Wiser* by Cass Sunstein and Reid Hastie, *The Second Machine Age* by Erik Brynjolfsson and Andrew McAfee, *Creative Conspiracy* by Leigh Thompson, *Darwin's Unfinished Symphony* by Kevin Laland, *Superforecasting* by

ACKNOWLEDGMENTS

Philip Tetlock and Dan Gardner, *Social Physics* by Alex Pentland, *Scale* by Geoffrey West, *Hit Refresh* by Satya Nadella, *From Bacteria to Bach and Back* by Daniel Dennett, and *The Logic of Scientific Discovery* by Karl Popper.

I would also like to thank the many people who agreed to be interviewed for the book, or who helped in other ways. These include Milo Jones, Geoffrey West, Carol Dweck, Jonathan Schulz, Duman Bahrami-Rad, Anita Woolley, Russell Lane, Satya Nadella, Matthew Stevenson, Michael Smith, Leigh Thompson, Yaya Fanusie, Ole Peters, Alex Adamou, Craig Knight, Eran Segal, and Jeffrey Mogford, and the wonderful staff at the Old Bank Hotel. Stuart Gent inspired the idea of using diagrams in chapter 2. Chapter 5 was largely influenced by the research of the brilliant philosopher C. Thi Nguyen and the psychologist Angela Bahns, as well as the book *Rising Out of Hatred* by Eli Saslow.

Above all, I would like to thank Kathy, my wife, Evie and Teddy, my children, and Abbas and Dilys, my parents. You are the best.

PERMISSIONS

NOTES

1: COLLECTIVE BLINDNESS

1 **On August 9, 2001:** Sources vary slightly on the precise date of Moussaoui's enrollment at the flight academy. I have used chapter 4 from *A Review of the FBI's Handling of Intelligence Information Related to the September 11 Attacks*, Office of the Inspector General, November 2004, https://oig.justice.gov/special/s0606/chapter4.htm.

1 **He paid for the bulk:** For more background see "How a 9/11 Conspirator Gave Himself Away," CNN.com, March 2, 2006, http://edition.cnn.com/2006/US/03/02/moussaoui.school/index.html.

3 **It was an outlier:** Bruce Hoffman, "The Modern Terrorist Mindset," in *Terrorism and Counterterrorism: Understanding the New Security*, eds. R. D. Howard and R. L. Sawyer (McGraw-Hill, 2011). See also Milo Jones and Philippe

Silberzahn, *Constructing Cassandra: Reframing Intelligence Failure at the CIA, 1947–2001* (Stanford Security Studies, 2013).

3 **In 1994, an Algerian group:** *The 9/11 Commission Report: Final Report of the National Commission on Terrorist Attacks Upon the United States* (W. W. Norton, 2004).

4 **On March 7, 2001:** "Russian Files on Al Qaeda Ignored," *Jane's Intelligence Digest* (October 5, 2001).

5 **He called it "creeping determinism":** Baruch Fischhoff and Ruth Beyth, "I Knew It Would Happen: Remembered Probabilities of Once-Future Things," *Organizational Behavior and Human Performance* 13 (February 1, 1975): 1–16, https://doi.org/10.1016/0030-5073(75)90002-1.

5 **Was the CIA being condemned:** Malcolm Gladwell, "Connecting the Dots: The Paradoxes of Intelligence Reform," *New Yorker* (March 10, 2003).

5 **As one counterterrorism:** Amy B. Zegart, *Spying Blind: The CIA, the FBI, and the Origins of 9/11* (Princeton University Press, 2009).

6 **When the CIA talked:** Author interview with anonymous source.

6 **In their meticulous study:** Jones and Silberzahn, *Constructing Cassandra.*

6 **"In 1964, the Office of National":** "The Case for Cultural Diversity in the Intelligence Community," *International Journal of Intelligence and Counter-Intelligence,* October 29, 2010, https://www.tandfonline.com/doi/pdf/10.1080/08850600150501317?needAccess=true.

7 **"Nothing really changed":** Author interview.

7 **Talking about his experience:** Robert Gates, *From the Shadows: The Ultimate Insider's Story of Five Presidents and How They Won the Cold War* (Simon & Schuster, 2008).

7 **At a conference in 1999:** Jones and Silberzahn, *Constructing Cassandra.*

7 **There are no publicly available:** Jones and Silberzahn, *Constructing Cassandra.*

14 **The Japanese, on the other hand:** Richard Nisbett with Hyun-Jung Lee and Michael Muthukrishna, "Culture and Intelligence," April 12, 2016, https://www.youtube.com/watch?v=SbgNSk95Vkk.

15 **"Suppose you are a doctor":** See Mary L. Gick and Keith J. Holyoak, "Analogical Problem Solving," *Cognitive Psychology* 12 (1980): 306–355, http://reasoninglab.psych.ucla.edu/KH%20pdfs/Gick-Holy oak%2819 80%29Analogical%20Problem%20Solving.pdf.

16 **"A fortress was situated":** Gick and Holyoak, "Analogical Problem Solving."

18 **"An uncomfortable truth dawned":** Reni Eddo-Lodge, *Why I'm No Longer Talking to White People About Race* (Bloomsbury, 2017).

20 **A study by Chad Sparber:** Chad Sparber, "Racial Diversity and Aggregate

Productivity," *Southern Economic Journal* 75, no. 3 (January 2009): 829–856; Richard Florida and Gary Gates, "Technology and Tolerance: The Importance of Diversity to High-Tech Growth," *Research in Urban Policy* 9 (December 2003): 199–219.

20 **A McKinsey analysis:** For French companies, the difference in return on investment was not significant.

22 **But here is an insider:** Quoted in Philip Shenon, *The Commission: The Uncensored History of the 9/11 Investigation* (Twelve, 2008).

22 **"How can a man in a cave":** *The 9/11 Commission Report.*

22 **"They simply couldn't square":** Author interview.

23 **This was corroborated by a CIA insider:** Michael Scheuer, *Through Our Enemies' Eyes: Osama bin Laden, Radical Islam, and the Future of America* (Potomac Books, 2003).

23 **Muslims know, too, that Muhammad's:** Jones and Silberzahn, *Constructing Cassandra.*

24 **"The CIA couldn't perceive":** John Miller and colleagues, who wrote *The Cell: Inside the 9/11 Plot* (Hyperion, 2002), make the same point with arch understatement: the CIA "continued to overlook, or at least underestimate, the breadth and power of the fundamentalist Islamic reform movement sweeping the Middle East."

27 **This is their prize:** The one unit specifically tasked with monitoring Al Qaeda was shunted out to a facility in Northern Virginia and its head person marginalized. When he sent a warning to the head of the CIA, he was effectively demoted to junior librarian.

27 **"It would be a mistake to redefine":** Pillar was referencing nuclear, bacterial, and chemical attacks, but as Jones and Silberzahn point out in *Constructing Cassandra*, Pillar "failed to appreciate . . . the possibility that 'grand' terrorism could be achieved with conventional approaches." See also Paul Pillar, *Terrorism and U.S. Foreign Policy* (Brookings Institution Press, 2003).

28 **Their men carried with them:** Other sources place bin Laden in Pakistan at the time of the attacks.

28 **A few minutes later:** Much of the chronology of the actions of the 9/11 terrorists taken from *Der Spiegel, Inside 9-11: What Really Happened* (St. Martin's Press, 2002) and Lawrence Wright, *The Looming Tower: Al Qaeda's Road to 9/11* (Penguin, 2007).

31 **"The methods for detecting":** *The 9/11 Commission Report.*

32 **But while it would be wrong:** There was too much emphasis on consensus and not enough on dissent. In her book *Spying Blind*, Zegart identifies structural

weakness at the agency. Various other concerns have been raised by scholars and broadly acknowledged by the CIA.

2: REBELS VERSUS CLONES

38 **"I became quite good"**: Michael Smith, *The Secrets of Station X: How the Bletchley Park Codebreakers Helped Win the War* (Biteback Publishing, 2011).

38 **"Imagine my surprise"**: Smith, *The Secrets of Station X.*

42 **Plato notes in *Phaedrus***: See Miller McPherson, Lynn Smith-Lovin, and James M Cook, "Birds of a Feather: Homophily in Social Networks," *Annual Review of Sociology* 27, no. 1 (August 2001): 415–44. http://aris.ss.uci.edu/~lin/52.pdf.

43 **"More than two decades later"**: Anthony King and Ivor Crewe, *The Blunders of Our Governments* (Oneworld, 2013).

46 **"Everyone projects onto others"**: King and Crewe, *The Blunders of Our Governments.*

47 **"She told me, shaking her head"**: Shane Snow, "Forget Culture Fit. Your Team Needs Culture Add.," LinkedIn.com, February 27, 2018, https://www.linkedin.com/pulse/forget-culture-fit-your-team-needs-add-shane-snow.

53 **"Each person's guess"**: James Surowiecki, *The Wisdom of the Crowds: Why the Many Are Smarter Than the Few* (Abacus, 2005).

55 **"Minority viewpoints are"**: Charlan Nemeth, "The Differential Contributions of Majority and Minority Influence," *Psychological Review* 93 (January 1, 1986): 23–32, https://www.researchgate.net/publication/232513627_The_Differential_Contributions_of_Majority_and_Minority_Influence.

57 **"Suppose you are building"**: Scott E. Page, *The Diversity Bonus: How Great Teams Pay Off in the Knowledge Economy* (Princeton University Press, 2017).

60 **"I stumbled on a counterintuitive"**: Scott E. Page, *The Difference: How the Power of Diversity Creates Better Groups, Firms, Schools, and Societies* (Princeton University Press, 2007).

61 **One survey of sports**: "Women in Sports Are Often Underrepresented in Science," *Science News* (blog), May 25, 2016, https://www.sciencenews.org/blog/scicurious/women-sports-are-often-underrepresented-science.

63 **Denniston cast his net**: Robin Denniston, *Thirty Secret Years: A. G. Denniston's Work in Signals Intelligence 1914–1944* (Polperro Heritage Press, 2007).

64 **Cryptography's loss**: Smith, *The Secrets of Station X.*

64 **There were many high-ranking Jewish**: Sinclair McKay, *The Secret Life of Bletchley Park: The History of the Wartime Codebreaking Centre by the Men and Women Who Were There* (Aurum Press, 2010).

64 **They were called Cillies:** Smith, *The Secrets of Station X.*

64 **"I am the world's expert":** Smith, *The Secrets of Station X.*

65 **"Crosswords are the same":** Sinclair McKay, "Could You Have Been a Bletchley Park Codebreaker? Take These Fiendish Puzzles and Solve This Crossword in 12 Minutes to See If You Might Have Been Recruited to Crack the Enigma Code," *Daily Mail*, October 13, 2017, https://www.dailymail .co.uk/news/article-4979048/Could-Bletchley-Park-codebreaker.html.

65 **"Crosswords are about getting":** McKay, "Could You Have Been a Bletchley Park Codebreaker?"

66 **Bill Bundy, an American:** Smith, *The Secrets of Station X.*

3: CONSTRUCTIVE DISSENT

70 **"I asked Jan":** Jon Krakauer, *Into Thin Air: A Personal Account of the Mt. Everest Disaster* (Macmillan, 1997).

70 **Their first date:** Shanee Edwards, "Everest: Rob Hall's Wife Jan Arnold Shares Her Story of Loss & Fear," SheKnows, January 26, 2016, https://www .sheknows.com/entertainment/articles/1109945/interview-jan-arnold-rob -halls-wife-everest/.

70 **"Brain cells were dying":** Krakauer, *Into Thin Air.*

71 **"The rock itself":** Edmund Hillary, *The View from the Summit* (Transworld, 1999).

71 **Since the mountain had first:** These figures apply to the time of the 1996 Everest expedition.

73 **"With enough determination":** Krakauer, *Into Thin Air.*

73 **"My altimeter read":** Krakauer, *Into Thin Air.*

74 **Pitman, who spent years haunted:** Ben Hoyle, "Everest Film Assassinates My Character, Says Climber," *The Times,* September 24, 2015, https://www .thetimes.co.uk/article/everest-film-assassinates-my-character-says-climber -87frkp3j87z.

76 **"The human mind is":** Jon Maner and Charleen Case, "Dominance and Prestige: Dual Strategies for Navigating Social Hierarchies," *Advances in Experimental Social Psychology* 54 (March 10, 2016): 129–180, https://www .researchgate.net/publication/297918722_Dominance_and_Prestige_Dual _Strategies_for_Navigating_Social_Hierarchies.

79 **According to the National:** A point made by Malcolm Gladwell in his book *Outliers: The Story of Success* (Little, Brown and Company, 2008).

79 **In one wide-ranging analysis:** Ayako Okuyama, Cordula Wagner, and Bart Bijnen, "Speaking Up for Patient Safety by Hospital-Based Health Care

Professionals: A Literature Review," *BMC Health Services Research* 14 (February 8, 2014): 61, https://www.ncbi.nlm.nih.gov/pubmed/24507747.

79 **"Actually, if you're a psychologist"**: "How Speaking Up Can Save Lives," *BBC News*, 26 July 2015. https://www.bbc.co.uk/news/health-33544778.

79 **"One psychologist monitoring"**: "How Speaking Up Can Save Lives."

80 **A clever study**: Balazs Szatmari, "We Are (All) the Champions: The Effect of Status in the Implementation of Innovations," Erasmus Research Institute of Management, December 16, 2016, https://repub.eur.nl/pub/94633/.

80 **"I believe that this happens"**: David Silverberg, "Why You Need to Question Your Hippo Boss," *BBC News*, April 19, 2017, https://www.bbc.co.uk/news/business-39633499.

80 **"HiPPOs rule the world"**: Avinash Kaushik, "Seven Steps to Creating a Data Driven Decision Making Culture," *Occam's Razor* (blog), October 23, 2006, https://www.kaushik.net/avinash/seven-steps-to-creating-a-data-driven-decision-making-culture/.

83 **"Yeah, it's turned out"**: Krakauer, *Into Thin Air*.

83 **"I felt like a part"**: *Storm Over Everest*, a film by David Breashears.

84 **"I was definitely considered"**: Krakauer, *Into Thin Air*.

84 **"[He gave] us a lecture"**: Krakauer, *Into Thin Air*.

88 **"And now the noise"**: Breashears, *Storm Over Everest*.

89 **"You just suck yourself"**: Breashears, *Storm Over Everest*.

90 **"It must have been a desperate struggle"**: Breashears, *Storm Over Everest*.

91 **"Meetings predict terrible"**: Author interview.

92 **"They are adamant"**: Author interview. And see also Leigh Thompson, *Creative Conspiracy: The New Rules of Breakthrough Collaboration* (Harvard Business Review Press, 2013).

92 **"Having really smart people"**: Author interview.

95 **"Much of the time"**: Cass Sunstein and Reid Hastie, *Wiser: Getting Beyond Groupthink to Make Groups Smarter* (Harvard Business Review Press, 2014).

95 **"Early on, founders Larry"**: Adam Galinsky and Maurice Schweitzer, *Friend and Foe: When to Cooperate, When to Compete, and How to Succeed at Both* (Crown, 2015).

96 **The takeaway was clear**: Eric Anicich et al., "The Costs of Co-Leadership in Fashion Houses, Mountaineering Teams, Qualitative Reports, and the Lab," *Academy of Management Proceedings* 2017, no. 1 (August 1, 2017): 11655, https://journals.aom.org/doi/10.5465/ambpp.2017.313.

96 **"Coleadership can kill"**: Galinsky and Schweitzer, *Friend and Foe*.

96 **"There is another important"**: Quoted in Joseph Henrich, *The Secret of Our Success* (Princeton University Press, 2015).

97 **To distinguish this form:** J. Henrich and F. J. Gil-White, "The Evolution of Prestige: Freely Conferred Deference as a Mechanism for Enhancing the Benefits of Cultural Transmission," *Evolution and Human Behavior: Official Journal of the Human Behavior and Evolution Society* 22, no. 3 (May 2001): 165–96, https://www.ncbi.nlm.nih.gov/pubmed/11384884.

97 **"Both dominance and prestige":** Author interview.

99 **"rank competitions":** J. K. Maner and C. R. Case, "Dominance and Prestige," *Advances in Experimental Social Psychology,* vol. 54 (Elsevier, 2016), 129–80, https://static1.squarespace.com/static/56cf3dd4b6aa60904403973f/t/57be 0776f7e0ab26d736060e/1472071543508/dominance-and-prestige-dual-strategies-for-navigating-social-hierarchies.pdf.

100 **"This boosts collective":** Author interview.

100 **A major investigation:** M. Lance Frazier et al., "Psychological Safety: A Meta-Analytic Review and Extension," *Personnel Psychology* 70, no. 1 (March 1, 2017): 113–65, https://creighton.pure.elsevier.com/en/publications /psychological-safety-a-meta-analytic-review-and-extension.

100 **"And it affects pretty much":** "The Five Keys to a Successful Google Team," re:Work, November 17, 2015, https://rework.withgoogle.com/blog/five-keys -to-a-successful-google-team/.

101 **"They are wrong":** Author interview.

101 **"There is a time":** Author interview.

102 **"Where I run some risk":** "The Beauty of Amazon's 6-Pager," LinkedIn, September 22, 2015, https://www.linkedin.com/pulse/beauty-amazons-6- pager-brad-porter.

103 **"It means that you gain":** Author interview.

104 **"The greatest tragedy":** Quoted in Adam Grant, *Originals: How Non-Conformists Change the World* (W. H. Allen, 2017).

104 **The researchers were interested:** "Hierarchical Cultural Values Predict Success and Mortality in High-Stakes Teams," *Proceedings of the National Academy of Sciences* 112, no. 5 (February 2015): 1338–1343, https://doi.org/10.1073 /pnas.1408800112.

105 **"In cultures that are hierarchical":** Galinsky and Schweitzer, *Friend and Foe.*

105 **"The Himalayan context":** Galinsky and Schweitzer, *Friend and Foe.* Interestingly, while hierarchy led to more fatalities, it also led to more climbers reaching the summit. Why? When I spoke to Anicich, the lead researcher, he said that it hinged on context. Dominance works when the conditions are stable and teamwork is mostly about coordination and speed. When conditions are complex and changing, however, dominance shades into danger. "This is when a leader needs to hear the perspectives of the team," he said.

106 **A study led by Michele:** Michele Gelfand et al., "Differences Between Tight and Loose Cultures: A 33-Nation Study," *Science* 332 (May 27, 2011): 1100–1104, https://www.researchgate.net/publication/51169484_Differences_Between _Tight_and_Loose_Cultures_A_33-Nation_Study.

106 **The researchers then divided:** Stephen Sales, "Economic Threat as a Determinant of Conversion Rates in Authoritarian and Nonauthoritarian Churches," *Journal of Personality and Social Psychology* 23, no. 3 (September 1972): 420–428.

108 **"Crippled by frostbite":** Stephen Venables, "Obituary: Rob Hall," *The Independent,* October 23, 2011, https://www.independent.co.uk/news/obituaries/obituary -rob-hall-1348607.html.

109 **"I love you":** It is likely that Harris and Hansen were dead when Hall spoke his last words. See also Krakauer, *Into Thin Air.*

4: INNOVATION

113 **"a group's ability":** Ian Morris, *Why the West Rules—For Now: The Patterns of History and What They Reveal About the Future* (Profile, 2011).

113 **"The Industrial Revolution ushered":** Erik Brynjolfsson and Andrew McAfee, *The Second Machine Age: Work, Progress, and Prosperity in a Time of Brilliant Technologies* (W. W. Norton, 2014).

114 **The production process:** Brynjolfsson and McAfee, *The Second Machine Age.*

114 **"Today, of course":** Andrew McAfee and Erik Brynjolfsson, *Machine, Platform, Crowd: Harnessing Our Digital Future* (W. W. Norton, 2017).

115 **The economist Shaw Livermore:** Shaw Livermore, "The Success of Industrial Mergers," *Quarterly Journal of Economics* 50, no. 1 (November 1935): 68–96.

115 **It was one of the most brutal:** Richard Caves, Michael Fortunato, and Pankaj Ghemawat, "The Decline of Dominant Firms, 1905–1929," *Quarterly Journal of Economics* 99 (February 1, 1984): 523–46, https://www.researchgate.net /publication/24092915_The_Decline_of_Dominant_Firms_1905—1929.

115 **"Everybody I took it to":** Scott Mayerowitz, "The Suitcase with Wheels Turns 40," ABC News, October 1, 2010, https://abcnews.go.com/Travel/suitcase -wheels-turns-40-radical-idea-now-travel/story?id=11779469.

116 **"In the first decades":** McAfee and Brynjolfsson, *Machine, Platform, Crowd.*

117 **"Sex is what makes":** Matt Ridley, *The Rational Optimist: How Prosperity Evolves* (4th Estate, 2010).

118 **He looked at 17.9:** "A Virtuous Mix Allows Innovation to Thrive," *Kellogg Insight,* November 4, 2013, https://insight.kellogg.northwestern.edu/article /a_virtuous_mix_allows_innovation_to_thrive.

119 **"Many of these novel":** "A Virtuous Mix."

119 **The vast majority of patents:** Hyejin Youn, Deborah Strumsky, Luis M. A. Bettencourt, and José Lobo, "Invention as a Combinatorial Process: Evidence from US Patents," *Journal of the Royal Society Interface* 12, no. 106 (May 2015), http://doi.org/10.1098/rsif.2015.0272.

119 **"The data reveal":** Page, *The Diversity Bonus.* See Brynjolfsson and McAfee, *The Second Machine Age.*

120 **"Each development becomes":** Brynjolfsson and McAfee, *The Second Machine Age.*

121 **These companies produced:** "Immigrant Founders of the 2017 Fortune 500," Center for American Entrepreneurship, 2017, http://startupsusa.org /fortune500/.

121 **More than half:** Sari Pekkala Kerr et al., "Global Talent Flows," *Journal of Economic Perspectives* 30, no. 4 (November 1, 2016): 83–106, https://pubs .aeaweb.org/doi/pdfplus/10.1257/jep.30.4.83.

121 **Different studies have shown:** Dane Stangler and Jason Wiens, "The Economic Case for Welcoming Immigrant Entrepreneurs," Ewing Marion Kauffman Foundation, March 26, 2014, https://www.kauffman.org/what -we-do/resources/entrepreneurship-policy-digest/the-economic-case-for -welcoming-immigrant-entrepreneurs.

121 **Another study, this time by Harvard Business School:** Harvard Business School, https://www.hbs.edu/faculty/Publication%20Files/17–011_da2c1cf4 -a999–4159-ab95–457c783e3fff.pdf.

121 **Yet another showed:** "The Economic Case."

122 **"It is exactly because":** McAfee and Brynjolfsson, *Machine, Platform, Crowd.*

123 **Instead of players:** See also Erik Dane, "Reconsidering the Trade-off Between Expertise and Flexibility," *Academy of Management Review* 35, no. 4 (October 2010): 579–603.

124 **Indeed, those who stayed:** Peter Vandor and Nikolaus Franke, "See Paris and . . . Found a Business? The Impact of Cross-Cultural Experience on Opportunity Recognition Capabilities," *Journal of Business Venturing* 31, no. 4 (July 1, 2016): 388–407, https://www.sciencedirect.com/science/article/pii /S0883902616300052.

124 **Those who imagined:** William W. Maddux and Adam D. Galinsky, "Cultural Borders and Mental Barriers: The Relationship Between Living Abroad and Creativity," *Journal of Personality and Social Psychology* 96, no. 5 (2009): 1047–61, https://www.apa.org/pubs/journals/releases/psp9651047 .pdf.

125 **One study found:** Robert S. Root-Bernstein, Maurine Bernstein, and Helen Gamier, "Identification of Scientists Making Long-Term, High-Impact Contributions, with Notes on Their Methods of Working," *Creativity Research Journal* 6, no. 4 (January 1993): 329–43, https://www.squawkpoint.com/wp-content /uploads/2017/01/Identification-of-scientists-making-long%E2%80%90term -high%E2%80%90impact-contributions-with-notes-on-their-methods-of -working.pdf.

125 **The Nobel laureates:** Robert Root-Bernstein et al., "Arts Foster Scientfic Success: Avocations of Nobel, National Academy, Royal Society, and Sigma Xi Members," *Journal of Psychology of Science and Technology* 1, no. 2 (2008): 51–63, https://www.psychologytoday.com/files/attachments/1035/arts-foster-scientific -success.pdf.

126 **Possibilities and opportunities:** Catherine Wines, "Why Immigrants Are Natural Entrepreneurs," *Forbes,* September 7, 2018, https://www.forbes.com/sites /catherinewines/2018/09/07/why-immigrants-are-natural-entrepreneurs/.

127 **"Sometimes (often actually)":** Jeff Bezos, "2018 Letter to Shareholders," *Day One: The Amazon Blog,* April 11, 2019, https://blog.aboutamazon.co.uk /company-news/2018-letter-to-shareholders.

129 **"The thing about ideas":** Gemma Corrigan, "How Can We Make Growth Work for All?," World Economic Forum, November 22, 2016, https://www.weforum .org/agenda/2016/11/introducing-a-new-competition-to-crowdsource-a -more-inclusive-economy/.

129 **Not only was the innovation:** Ridley, *The Rational Optimist.*

130 **They also needed sophisticated:** See Steven Johnson, *Where Good Ideas Come From: The Seven Patterns of Innovation* (Riverhead, 2010).

130 **For those interested in existentialism:** Randall Collins, *The Sociology of Philosophies: A Global Theory of Intellectual Change* (Belknap Press, 1998).

131 **"Intellectual creativity":** Collins, *The Sociology of Philosophies.*

133 **"A common perception":** Michael Muthukrishna and Joseph Henrich, "Innovation in the Collective Brain," *Philosophical Transactions of the Royal Society B: Biological Sciences* 371, no. 1690 (March 19, 2016): 20150192, https:// royalsocietypublishing.org/doi/full/10.1098/rstb.2015.0192.

134 **"Sunspots were simultaneously":** Johnson, *Where Good Ideas Come From.*

134 **Bigger networks permitted:** Michelle A. Kline and Robert Boyd, "Population Size Predicts Technological Complexity in Oceania," Proceedings of the Royal Society B: Biological Sciences 277, no. 1693 (April 14, 2010): 2559–2564, https://royalsocietypublishing.org/doi/full/10.1098/rspb.2010.0452.

135 **As the Pama-Nyungan:** Henrich, *The Secret of Our Success.*

135–136 **The populations were genetically:** Henrich, *The Secret of Our Success.*

137 **Among the geniuses:** Joseph Henrich and Michael Muthukrishna argue that differences in individual IQ are an emergent property of the collective brain. See Joseph Henrich and Michael Muthukrishna, "Innovation in the Collective Brain."

137 **"If you want to have cool technology":** Henrich, *The Secret of Our Success.*

138 **In 1975:** AnnaLee Saxenian, *Regional Advantage: Culture and Competition in Silicon Valley and Route 128* (Harvard University Press, 1994).

138 **Land and office space:** Saxenian, *Regional Advantage.*

139 **"a sociological unit":** Glenn Rifkin and George Harrar, *The Ultimate Entrepreneur: The Story of Ken Olsen and Digital Equipment Corporation* (Contemporary Books, 1988).

139 **"Practices of secrecy":** Saxenian, *Regional Advantage.*

140 **"The technology companies":** Saxenian, *Regional Advantage.*

140 **"Every year there was some place":** Tom Wolfe, "The Tinkerings of Robert Noyce: How the Sun Rose on the Silicon Valley," *Esquire* (December 1983).

141 **"Are you building":** Walter Isaacson, *Innovators: How a Group of Inventors, Hackers, Geniuses, and Geeks Created the Digital Revolution* (Simon & Schuster, 2014).

142 **"It was a Eureka":** Jessica Dolcourt, "Apple's 40-Year Legacy Began with This 'Eureka' Moment," *CNET,* August 10, 2016, https://www.cnet.com/news/steve-wozniak-on-homebrew-computer-club/.

142 **"I was not aware":** Saxenian, *Regional Advantage.*

144 **"Some of my most sobering":** James Temple, "Tech's Lost Chapter: An Oral History of Boston's Rise and Fall," *Vox,* December 9, 2014, https://www.vox.com/2014/12/9/11633606/techs-lost-chapter-an-oral-history-of-bostons-rise-and-fall-part-one.

144 **"There is a unique atmosphere":** Saxenian, *Regional Advantage.*

146 **"hastily constructed of plywood":** David Shaffer, "Building 20: What Made It So Special and Why It Will (Probably) Never Exist Again," *Daily Journal of Commerce,* June 19, 2012, http://djcoregon.com/news/2012/06/19/building-20-what-made-it-so-special-and-why-it-will-probably-never-exist-again/.

147 **"Who would have thought":** Another line of research is conducted by network theorists themselves. One famous study by Sandy Pentland of MIT analyzed eToro, a platform for financial traders. Users can look up each other's trades, portfolios, and past performance and can copy trading ideas if they think it will increase their own profits. Pentland and his colleagues collected data from 1.6 million users, tracking almost everything about the exchanges between them and financial return.

They found that traders who were isolated in the network performed poorly. They had "impoverished opportunities for social learning because they had too few links to others." But the researchers also found that people who were highly interconnected also performed poorly. Why? Because they were embedded in a web of feedback loops, so they were hearing the same ideas over and over again. They were caught up in echo chambers.

It was traders whose networks exposed them to new ideas, but not merely recycled stale ideas, who performed the best. In fact, by subtly reshaping the structure of the network, and by offering small incentives to nudge people out of the echo chambers, Pentland was able to raise the financial return of the entire network. "By reducing idea flow to allow greater diversity, we moved the social network back into its sweet spot and raised average performance," he said.

147 **You can see the power of networks:** These points about recombinant innovation in soccer were also made in my column for *The Times*: Matthew Syed, "Reluctance to Embrace 'Idea Sex' Is Stopping English Football from Evolving," *The Times,* April 2, 2018, https://www.thetimes.co.uk/article/why-english-footballs-reluctance-to-embrace-idea-sex-is-stopping-the-game-from-evolving-gs75vb30v.

149 **"It wasn't necessarily":** Owen Slot, *The Talent Lab: How to Turn Potential into World-Beating Success* (Ebury, 2017).

149 **"F1 technology":** Slot, *The Talent Lab.*

150 **"It was not a business":** "Science in the Scottish Enlightenment: 3.1 Clubs and Societies," OpenLearn, 2020, https://www.open.edu/openlearn/history-the-arts/history/history-science-technology-and-medicine/science-the-scottish-enlightenment/content-section-3.1.

150 **"The interconnections and cross-fertilization":** "Science in the Scottish Enlightenment."

5: ECHO CHAMBERS

153 **"the most visited white supremacist":** Tara McKelvey, "Father and Son Team on Hate Site," *USA Today,* July 16, 2001, https://usatoday30.usatoday.com/life/2001–07–16-kid-hate-sites.html.

154 **One study showed:** "White Homicide Worldwide," Southern Poverty Law Center, March 31, 2014, https://www.splcenter.org/20140331/white-homicide-worldwide.

154 **He won a seat:** Matt Stieb, "Ex–White-Nationalist Says They Get Tips From Tucker Carlson," *The Intelligencer,* April 1, 2019, http://nymag.com/intelligencer/2019/04/ex-white-nationalist-says-they-get-tips-from-tucker-carlson.html.

156 **"I'd like to introduce":** See Eli Saslow, *Rising Out of Hatred: The Awakening of a Former White Nationalist* (Doubleday, 2018). Also see R. Derek Black, Khalil Muhammad, and Elle Reeve, "'I'm Not a Racist, But...': Examining the White Nationalist Efforts to Normalize Hate," The Institute of Politics at Harvard University, October 18, 2017, https://iop.harvard.edu/forum/im-not-racist -examining-white-nationalist-efforts-normalize-hate.

157 **This is a diverse population:** Data provided by the academic Angela Bahns, personal correspondence.

157 **In addition to the University of Kansas:** Angela J. Bahns et al., "Similarity in Relationships as Niche Construction: Choice, Stability, and Influence Within Dyads in a Free Choice Environment," *Journal of Personality and Social Psychology* 112, no. 2 (February 2017): 329–55, https://www.ncbi.nlm .nih.gov/pubmed/26828831.

158 **Bethel has just 105:** Data provided by Bahns, measured in 2009.

159 **"It sounds ironic":** Author interview.

159 **In one experiment led by:** Paul Ingram and Michael W. Morris, "Do People Mix at Mixers? Structure, Homophily, and the 'Life of the Party,'" *Administrative Science Quarterly*, 52 (2007): 558–585, http://www.columbia.edu/~pi 17/mixer.pdf.

162 **This is where various algorithms:** Eli Pariser, *The Filter Bubble: What the Internet Is Hiding from You* (Viking, 2011).

162 **"The red group says":** "See How Red Tweeters and Blue Tweeters Ignore Each Other on Ferguson," *Quartz*, November 25, 2014, https://qz.com/302616 /see-how-red-tweeters-and-blue-tweeters-ignore-each-other-on-ferguson/.

162n12 **A different study concluded:** Ana Lucia Smith, "A Quantitative Analysis of News Consumption on Facebook," IMT School for Advanced Studies, 2017, https://pdfs.semanticscholar.org/e05f/05f773c9fc3626fa20f9270e6cefd899 50db.pdf and https://arxiv.org/abs/1903.00699.

164 **It was as if exposure:** Christopher A. Bail et al., "Exposure to Opposing Views on Social Media Can Increase Political Polarization," *Proceedings of the National Academy of Sciences of the United States of America* 115, no. 37 (September 11, 2018): 9216–21, https://www.ncbi.nlm.nih.gov/pmc/articles /PMC6140520/.

164 **Echo chambers, Nguyen argues:** Elizabeth Dubois and Grant Blank, "The Echo Chamber Is Overstated: The Moderating Effect of Political Interest and Diverse Media," *Information, Communication & Society* 21, no. 5 (May 4, 2018): 729–45, https://www.tandfonline.com/doi/pdf/10.1080/1369118X.2018. 1428656.

164 **In their scholarly book:** Kathleen Hall Jamieson and Joseph N. Cappella, *Echo Chamber: Rush Limbaugh and the Conservative Media Establishment* (Oxford University Press, 2010).

166 **"What's happening is a kind":** C. Thi Nguyen, "Escape the Echo Chamber," *Aeon* (April 9, 2018), https://aeon.co/essays/why-its-as-hard-to-escape-an -echo-chamber-as-it-is-to-flee-a-cult.

166 **"Ask yourself":** Nguyen, "Escape the Echo Chamber."

167 **"The world of anti-vaccination":** Nguyen, "Escape the Echo Chamber."

167 **"Here's a basic check":** Nguyen, "Escape the Echo Chamber."

167 **"Derek was socialized":** Saslow, *Rising Out of Hatred.*

168 **"He was impervious":** For much of the biographical detail in this section, see Saslow, *Rising Out of Hatred.*

176 **"A large section":** "Derek Black email to Mark Potok," Southern Poverty Law Center, July 15, 2013, https://www.splcenter.org/sites/default/files/derek -black-letter-to-mark-potok-hatewtach.pdf.

179 **A paper by the Finnish:** Jaakko Hintikka, "The fallacy of fallacies," *Argumentation* 1, no. 3 (1987): 211–238, https://philpapers.org/rec/HINTFO-3.

179 **"It may be worth":** John Locke, *An Essay Concerning Human Understanding* (Clarendon Press, 1975).

6: BEYOND AVERAGE

184 **"It wasn't what I expected":** Material on Eran and Keren Segal taken from a personal interview and Eran Segal and Eran Elinav, *The Personalized Diet: The Revolutionary Plan to Help You Lose Weight, Prevent Disease and Feel Incredible* (Vermilion, 2017).

188 **"You never knew":** Todd Rose, *The End of Average: How to Succeed in a World That Values Sameness* (Penguin, 2017).

188 **In total, 172 incidents:** "USAF INFORMATION for 1950," AccidentReport .com, 2020, http://www.accident-report.com/Yearly/1950/5002.html.

189 **"thumb length, crotch height":** Rose, *The End of Average.*

190 **Even when Daniels:** Rose, *The End of Average.*

192 **The single IQ metric:** Rose, *The End of Average.*

193 **After all, these are the admin:** Amy Wrzesniewski et al., "Dual Mindsets at Work: Achieving Long-Term Gains in Happiness" (working paper, 2017).

193 **"We introduced hundreds":** Grant, *Originals.*

195 **"When it's used badly":** Author interview.

196 **"A good example is the so-called":** Author interview.

NOTES

199 **These subjects were then connected:** David Zeevi et al., "Personalized Nutrition by Prediction of Glycemic Responses," *Cell* 163, no. 5 (November 19, 2015): 1079–94, https://www.ncbi.nlm.nih.gov/pubmed/26590418.

199 **"It was sobering":** Detail from this chapter taken from interviews with Segal and others, plus Segal and Elinav, *The Personalized Diet.*

204 **"We were initially shocked":** Author interview.

206 **"standardized curricula":** Todd Rose and Ogi Ogas, *Dark Horse: Achieving Success Through the Pursuit of Fulfillment* (HarperOne, 2018).

207 **"Our schools are, in a sense":** Ellwood Cubberley, *Public School Administration: A Statement of the Fundamental Principles Underlying the Organization and Administration of Public Education* (1916).

207 **Five of these factors:** "6 Key Principles That Make Finnish Education a Success," *EdSurge*, July 31, 2018, https://www.edsurge.com/news/2018-07-31-6-key-principles-that-make-finnish-education-a-success.

208 **In her book *Invisible Women*:** Caroline Criado Perez, *Invisible Women: Exposing Data Bias in a World Designed for Men* (Abrams Press, 2019).

209 **They were then asked to press:** Michael B. Miller et al., "Extensive Individual Differences in Brain Activations Associated with Episodic Retrieval Are Reliable over Time," *Journal of Cognitive Neuroscience* 14, no. 8 (November 15, 2002): 1200–1214, https://www.ncbi.nlm.nih.gov/pubmed/12495526.

209 **"Most didn't look":** Rose, *The End of Average.*

210 **"It was called the lean office":** Author interview.

211 **Subjects had to check:** "JEP Space Experiments," Wordpress, July 2013, https://adobe99u.files.wordpress.com/2013/07/2010+jep+space+experiments.pdf.

214 **Further research is taking place:** Some of the most recent is led by Tim Spector, an epidemiologist at King's College, London.

7: THE BIG PICTURE

220 **"Once population size":** Kevin N. Laland, *Darwin's Unfinished Symphony: How Culture Made the Human Mind* (Princeton University Press, 2017).

222 **"Techniques such as cooking":** Author interview. See also Henrich, *The Secret of Our Success.*

226 **"Humanity's success is sometimes":** Laland, *Darwin's Unfinished Symphony.*

230 **In her book *What Works*:** Another way of removing bias is by using algorithms to make hiring decisions or, at the very least, to whittle down the list of candidates. After all, machines are not subject to the stereotyping that often influences human judgment. At least, that is the theory.

In truth, as the author Cathy O'Neil has shown in her book *Weapons of Math Destruction* (Penguin, 2017), algorithms can themselves reflect the biases that exist within societies. She relates the case of Gild, an American start-up that looks at millions of data points to assess the suitability of candidates for jobs, mainly in the tech industry. One predictor of job success is how well integrated a coder is with the coding community. Those with larger followings score higher, as do those connected to influential coders.

But while seeking correlations, the Gild algorithm finds other patterns, too. It turns out, for example, that frequenting a Japanese manga site is a "solid predictor of strong coding." On the surface, this sounds like a useful piece of information for any company hoping to recruit top coders.

Yet now consider the effects on gender. Women, on average, perform 75 percent of the world's unpaid care work. A talented female coder might therefore be expected, on average, to have less time to spend hours on manga and other sites. And if the content of the website is not women friendly, they are even less likely to visit it. As O'Neil puts it, "If, like most of techdom, that manga site is dominated by males and has a sexist tone, a good number of women in the industry will probably avoid it."

This means that an algorithm that lowers the relative score of people not visiting such sites will entrench an unfair bias against talented female coders. "Gild undoubtedly did not intend to create an algorithm that discriminated against women," writes Caroline Criado Perez. "They were intended to remove human biases. But if you aren't aware of how those biases operate, if you aren't collecting data and taking a little time to produce evidence-based processes, you will continue to blindly perpetuate old injustices. And so by not considering ways in which women's lives differ from men's, both on and offline, Gild's coders inadvertently created an algorithm with a hidden bias against women."

232 **In the same period, Prada's:** Jennifer Jordan and Michael Sorell, "Why You Should Create a 'Shadow Board' of Younger Employees," *Harvard Business Review,* June 4, 2019, https://hbr.org/2019/06/why-you-should-create-a -shadow-board-of-younger-employees.

240 **"He was by far":** Scott Shane, "Drone Strike That Killed Awlaki 'Did Not Silence Him,' Journalist Says," *Fresh Air,* September 14, 2015, https://www .npr.org/2015/09/14/440215976/journalist-says-the-drone-strike-that-killed -awlaki-did-not-silence-him.

INDEX

ABOUT THE AUTHOR

Matthew Syed is the *Sunday Times* number one bestselling author of *Bounce, Black Box Thinking,* and *You Are Awesome.* He writes an award-winning newspaper column in *The Times* and is the cohost of the hugely successful BBC podcast *Flintoff, Savage and the Ping Pong Guy.* Matthew is the cofounder of Greenhouse, a charity that empowers youngsters through sports, a member of the Football Association's Technical Advisory Board, and an ambassador for PixL, an educational foundation. Matthew lives in London with his wife and two children. To find out more about his work visit matthewsyed.co.uk.